ALIBUS

ALS AND CULTURES

ROTHFELS, GENERAL EDITOR

in the Animalibus series share a fascination with the status and le of animals in human life. Crossing the humanities and the social es to include work in history, anthropology, social and cultural geog-, environmental studies, and literary and art criticism, these books hat thinking about nonhuman animals can teach us about human res, about what it means to be human, and about how that meaning t shift across times and places.

ER TITLES IN THE SERIES:

el Poliquin, *The Breathless Zoo: lermy and the Cultures of Longing*

B. Landes, Paula Young Lee, and Youngquist, eds., *Gorgeous Beasts: ıal Bodies in Historical Perspective*

Emma Thorsen, Karen A. Rader, and n Dodd, eds., *Animals on Display: Creaturely in Museums, Zoos, and ıral History*

-Janine Morey, *Picturing Dogs, Seeing selves: Vintage American Photographs*

y Sanders Pollock, *Storytelling Apes: natology Narratives Past and Future*

rid H. Tague, *Animal Companions: Pets l Social Change in Eighteenth-Century tain*

ck Blau and Nigel Rothfels, *Elephant use*

arcus Baynes-Rock, *Among the Bone ters: Encounters with Hyenas in Harar*

Monica Mattfeld, *Becoming Centaur: Eighteenth-Century Masculinity and English Horsemanship*

Heather Swan, *Where Honeybees Thrive: Stories from the Field*

Karen Raber and Monica Mattfeld, eds., *Performing Animals: History, Agency, Theater*

J. Keri Cronin, *Art for Animals: Visual Culture and Animal Advocacy, 1870–1914*

Elizabeth Marshall Thomas, *The Hidden Life of Life: A Walk Through the Reaches of Time*

Elizabeth Young, *Pet Projects: Animal Fiction and Taxidermy in the Nineteenth-Century Archive*

CROC

NIGEL

Adviso
Steve
Susan
Garry
Kari V

Books
the ro
scienc
raphy
ask w
cultu
migh

OTH
Rach
Taxi
Joan
Paul
Ani
Liv
Ada
The
Nat
Ann
Ou
Ma
Pri
Ing
an
Br
Di
H
M
E

CROCODILE UNDONE

The Domestication of Australia's Fauna

MARCUS BAYNES-ROCK

Foreword by Agustín Fuentes

THE PENNSYLVANIA STATE UNIVERSITY PRESS
UNIVERSITY PARK, PENNSYLVANIA

Photo credits: Page xvi, courtesy of Wendy Tanner; pages 26, 52, 82, 112, 134, Marcus Baynes-Rock; page 160 (isolators for chicken work in the Small Animal Facility at CSIRO's Australian Animal Health Laboratory in Geelong), Frank Filippi, CSIRO, February 9, 2005, licensed under CC BY 3.0.

Library of Congress Cataloging-in-Publication Data

Names: Baynes-Rock, Marcus, 1965– author.
Title: Crocodile undone : the domestication of Australia's fauna / Marcus Baynes-Rock ; foreword by Augustin Fuentes.
Other titles: Animalibus.
Description: University Park, Pennsylvania : The Pennsylvania State University Press, [2020] | Series: Animalibus: of animals and cultures | Includes bibliographical references and index.
Summary: "Examines issues surrounding the domestication of wild animals and the disruption of traditional ecologies in Australia"—Provided by publisher.
Identifiers: LCCN 2019058675 | ISBN 9780271086194 (hardcover)
Subjects: LCSH: Domestication—Australia. | Ecology—Australia.
Classification: LCC SF55.A8 B39 2020 | DDC 636.00994—dc23
LC record available at https://lccn.loc.gov/2019058675

Printed in the United States of America
Published by The Pennsylvania State University Press,
University Park, PA 16802–1003

The Pennsylvania State University Press is a member of the Association of University Presses.

It is the policy of The Pennsylvania State University Press to use acid-free paper. Publications on uncoated stock satisfy the minimum requirements of American National Standard for Information Sciences—Permanence of Paper for Printed Library Material, ANSI Z39.48–1992.

For Deborah Bird Rose

1946–2018 . . .

CONTENTS

Foreword: Undoing the World? (ix)
AGUSTÍN FUENTES

Acknowledgments (xiii)

1. The Great Unmaking (1)
2. Dingoes (27)
3. Stingless Bees (53)
4. Crocodiles (83)
5. Emus (113)
6. Kangaroos (135)
7. Borderlands (161)

Notes (185)

Bibliography (197)

Index (209)

FOREWORD: UNDOING THE WORLD?

Many humans believe that the world should be exploited for our benefit, a perspective especially prominent in nations and peoples committed to the contemporary neoliberal economic model. While there are societies that place limits on destructive exploitation, and in some cases even practice sustainable processes for the use of the plants, animals, space, and land with which they share the planet, a majority of human societies, religions, political systems, and economic entities act as though they believe that humans have the right to use the world as they see fit. Today, in the twenty-first century, capitalist market economies have become dominant, and the human global population continues to increase at a colossal pace. The explosion in the extraction of biotic and abiotic resources has reached a scale and impact such that the only way humans could possibly continue in this vein is by having a confidence that the earth can sustain us regardless of what we do to it.

Such a belief is wholly false. We have the data.

The chemist Will Steffen and colleagues, in their now classic essay on the Anthropocene, tell us that "the human imprint on the global environment has now become so large and active that it rivals some of the great forces of Nature in its impact on the functioning of the Earth system."[1] In 2017, a letter signed by more than sixteen thousand scientists from 184 countries titled "World Scientists' Warning to Humanity: A Second Notice" identified the key global trends in species extinctions, ozone depletion, carbon dioxide emissions, oceanic dead zones, temperature increase, the loss of forest, freshwater, and saltwater resources, and human population growth, which all portend approaching global catastrophe. In 2017, I joined a group of thirty-one primatology colleagues in publishing an article titled "Impending Extinction Crisis of the World's Primates: Why Primates Matter," focusing specifically on our close cousins, the other primates, identifying every major extinction threat, population loss, and habitat crisis, and determining that 100 percent of the cause is us.[2]

The human commitment, investment, and devotion to drawing resources from our planet is not sustainable at our current pace or at our current level of global population. And it is, in part, particular beliefs in ourselves and our relations to the world that facilitated these processes of climate change and ecological disruption.

But then again, we humans are consummate niche constructors. We are ecosystem managers. It is our history, and future, to shape the world as it shapes us. With such endeavors come attendant ethical and practical responsibilities. We have the capacities to shape sustainable futures for ourselves and, by doing so, to shape sustainable processes for the ecosystems we are part of and the myriad other species with which we share them.

In order to do this, we have to seriously and effectively reflect on what we are doing and how we are doing it. The contemporary reality of our relations to all the others that share this planet requires that we shed traditional modes of explanation and understanding of what humans "do" with other species, and why. We need to develop novel, more entangled, and less myopic (less anthropic?) visions of where we are and where we are going. There is no better place to tackle this conundrum than by taking apart and putting back together that most human of all enterprises: domestication. And this is exactly what Marcus Baynes-Rock invites us to do. In this excellent narrative, Baynes-Rock takes us on a journey that is simultaneously visceral, intellectual, engaging, distressing, and compassionate, forcing readers to shed our restrictive lenses and recognize the undoing processes of what we so glibly call "domestication."

Humanity has been and is shaped by our caretaking, consumption, manipulation, and destruction of, and our compassion for, other beings. Human evolution is a multispecies endeavor, a current in which humans and our companion species have always been caught. A current sweeping countless others into the processes of human becoming—forcing them along for the ride, co-shaping our and their futures. This process of entanglement is more pervasive and broadly distributed in the Anthropocene than ever before. But this is not just a simple story of how humans bent other species to our will over the last ten thousand years, creating the tame and the wild. As Baynes-Rock so eloquently lays out for us, this is a story of undoing. A more terrifying and yet more plausible narrative.

Take, for example, the chicken. There are more than twenty billion domestic chickens (*Gallus gallus domesticus*) on the planet at the moment, and because of our consumption patterns as many as three times that number will be born and eaten over the course of the next few years. By standard measures (the passing of genes from generation to generation), chickens, with their billions-strong population, are incredibly evolutionary successful. But can we actually say they've achieved any evolutionary success as a species? Without massive anthropogenic sustenance and structuring, they, in their contemporary biological forms, could not maintain anything close to current population levels. Driving home this point, Carys E. Bennett and colleagues argue that the chicken is actually a robust signal of a human reconfigured biosphere, noting that "chickens, now unable to survive without human intervention, have a combined mass exceeding that of all other birds on Earth; this novel morphotype symbolizes the unprecedented human reconfiguration of the Earth's biosphere."[3] To use Baynes-Rock's motif, we've "undone" the chicken. I am sure that every reader can instantly come up with more than a handful of other examples in which this undoing has wrenched species from particular trajectories and placed them squarely into others, unstructuring and restructuring what it means to be that species. For Baynes-Rock, it is "that which is taken away in the course of enfolding animals into human societies that in fact defines domestication." I agree, and I am certain that such a perspective is needed if we want to better understand where we (all of us, not just humans) are heading.

Baynes-Rock lays out a compelling argument that the key to understanding domestication is seeing it as the separation of animals from their traditional ecologies and incorporating (or forcing) them into a human niche—creating anthropogenic feeding, reproductive, and social systems wherein the animals (and maybe the humans too) become somewhat undone from one reality and melded into another.

But not all species are so easily undone. Nor have most human groups spent so much time and effort in the undoing as many in the contemporary geopolitical landscape do. Understanding this is the key to redoing at least some parts of the world. To drive this home, we accompany Baynes-Rock in his homeland of Australia, where he demonstrates myriad not quite domestic relations and uses them as a frame to show what is possible, impossible, and maybe even hopeful in humanity's engagement with others. This is his challenge to thinking simply with, about, and through "domestication."

Traversing landscapes of humans, crocodiles, dingoes, stingless bees, emus, and kangaroos, Baynes-Rock illustrates a new wave of "domestication" that is not what we think it is. There are no villains, and no heroes. There are just people and animals, including the author, manipulating, interacting, reshaping, undoing, and redoing. The new wave of domestication is not a carefully orchestrated program of change; it is a haphazard process that carries people, animals, and ecologies into unfamiliar landscapes. It is what has been and is happening, and seeing such processes in this frame enables a new, and possibly hopeful, landscape.

Baynes-Rock ensures our attention, and offers both complication and enjoyment, as he invites the reader into his ethnography, his family, and his world. He takes us on a narrative journey that melds rigorous scholarship, innovative and engaging writing, and powerful storytelling, leading us to a shared conclusion. Undoing matters and redoing is possible, but not easy.

In this book, Baynes-Rock destroys a suite of dichotomies—"Culture is nature, mind is body, soul is body, human is animal"—enabling us to better understand that in this story of relations, "there is no good and evil, only roles within ecosystems." After reading this book, it becomes very difficult to see humans and the world as separate entities. The stories of multiple species in messy, entangled histories and presents give us a better glimpse of the reality wherein we are all enmeshed in a range of discordant tangles, some bringing us closer and others pushing us apart.

I can think of no better way to close this foreword than to quote from the end of the book to help the reader dive into the narrative with the understanding of where she will emerge: "We need to embrace complexity and learn to trust land, plants, and animals in ways that allow life to flow across landscapes, sometimes toward and sometimes away from us. Only then will we be able to reconnect, not only with the deep webs of complex ecologies, but with our humanity."

—AGUSTÍN FUENTES

ACKNOWLEDGMENTS

I must express my deepest thanks to Agustín Fuentes, whose enthusiasm, open-mindedness, belief, and support brought this project to fruition. This book is as much his as it is mine. Thanks also to Celia Deane-Drummond, who is equally responsible for making the project happen. Celia seems to find meaning in my work that would otherwise elude me. I am indebted to Natasha Fijn for her insights and inspiration and for her critical reading and suggestions. Nigel Rothfels reviewed early drafts and, as always, offered sage advice that only benefited the book. My thanks to Kendra Boileau for her constant encouragement and for letting me get away with so many grammatical crimes. And thanks to the folks at PSU Press for turning my words into something tangible and to Suzanne Wolk for the masterly editing. Thanks to my friends and colleagues who listened patiently while I presented my work. I like to think that you weren't just there for the free drinks and nibbles. I must also thank the staff at Notre Dame: Rebecca Artinian-Kaiser and Katie Zakas Rutledge at the Center for Theology, Science and Human Flourishing, and Michelle Thornton and Eileen Barany in the Department of Anthropology. The funding for this project came from a grant from the John Templeton Foundation. I very much appreciate that I had complete freedom to explore this subject and let the cards fall where they may. This is how science progresses.

In the course of my fieldwork I've encountered so many willing and helpful souls, and I'm grateful to them all. Margi Woulfe of Yanigurra Dingo Sanctuary shared with me so much of her time and knowledge and gave me helpful feedback on early draft chapters. John Cooper also shared much of his dingo knowledge and his lateral thinking on managing predation on a farm. Thanks to Gwen Thornton for time spent with the puppies and a thought-provoking interview. I'm also deeply thankful to Tye Kennedy for his insights, his time, and for the yummy bunya nuts. Thanks to Tim Heard for sharing his knowledge and ideas over a cup of tea on the back veranda. Also in Brisbane, Nick Powell showed me the life worlds of Australia's native bees and how not caring is a form of care. I'm indebted to Wendy Tanner for telegraphing useful bits of information. And thanks to the owners and staff at Bindara crocodile farm, who readily gave up their time to show me around and sit for interviews. I wasn't very clear in telling them what my research was about—because I had little idea myself—but they willingly helped out regardless. Thanks to Shane at Emu Bliss, who did likewise despite a full workload at the time, and to Terry Turner of the American Emu Association for emu stats. Thanks to Barry Watson and Terry Walker, who took time out to talk emus with me. And special thanks to all of the wonderful people of Bourke who guided my project. Among those, Fiona Garland of Local Land Services was especially helpful. Thanks to Ross Kemp for being so damn amenable. And lastly, thanks to all of the people not mentioned here, who in some form or other made my life a little easier while I was doing the research and writing of this book. I hope I've done you proud.

1.

The Great Unmaking

The husbandry system, whose forms underlie the foundations of modern thought, excludes wild nature as chaotic, other, and evil.
—PAUL SHEPARD, *The Tender Carnivore and the Sacred Game*

"Disestablished." That's the word they used when they shut down our research center. I was in the middle of my hyena study when the word came through that the Centre for Research on Social Inclusion (CRSI) was to be disestablished. I recall marveling at the setting of a new standard for euphemistic bullshit. Yet I was unthrilled. When the doors closed, they handed me over to the Sociology Department, and Debbie Rose, who was advising me, was farmed out to Anthropology. Detritus. Debbie was becoming increasingly disillusioned with academia, and the closure of our beloved CRSI seems to have pushed her through a door into a world of cynicism. Australian universities have been corporatized. No longer institutions of public good, they have been rendered by economists and politicians as places of individual investment in the "knowledge economy."[1] Yet it is not knowledge per se that is being valued and traded; it is money-making potentiality. Staff, students, and alumni have become "stakeholders," research is judged in terms of its immediate return on investment, critical thinking has taken a backseat to job readiness, and academics are under increasing pressure to justify their worth in terms that can be measured by an accountant. Within this context, Debbie wrote a paper for a special issue of *Text*, the modest journal of the Australasian Association of Writing Programs. The article is titled "Slowly." I especially like the abstract that Debbie wrote: "This paper argues against abstracts."[2] The remainder of the essay is Debbie's crying out from within a system that seeks to abstract and itemize people and their work in order to meet a bottom line. Hence one of the keywords: "university-zombies." Her message for the journal's readers is that there needs to be space for "slow" writing, for academics to engage in dialogic modes of writing that cannot be nurtured within a system of measures of outputs and rankings. It's a heartfelt cry, and all the more poignant in light of Debbie's disillusionment. But for me, the essay holds a much more important message with far broader implications.

"Slowly" is not merely an S.O.S., desperately scrawled on the top of a faculty edifice surrounded by the walking dead. Debbie's message is an urgent wake-up call to a species that threatens to undo itself and bring down with it countless other species in the process. Humans are niche constructors par excellence—we are actively creating the conditions of our own selection—and of late, our industrialized, urbanized niche, which is supposed to free us from nature's teeth and claws, threatens selective pressure on our species at a scale not unlike that of an asteroid impact.[3] If it weren't for the dire consequences we face, this would be a delicious irony: the very defenses erected to protect late modernity from the world that sustains it are crumbling in the face of cataclysmic processes of its own (un)making. Debbie calls this a "necessary feedback loop" in a world of connections that late modernity not only ignores but actively seeks to sever.[4] She shows us the mountain that stands between ontologies of connection and ontologies of separation like an inverted chasm. It's not a metaphorical mountain but the real thing in Cape Breton Island, Nova Scotia, pressing its weight upon the earth, exerting its particular gravitational pull. The mountain known to the Indigenous Mi'kmaq people as Kluskap's Mountain was targeted by a local corporation called Kelly Rock. The directors of the company saw in the mountain a bounty of granite. Their plan was to strip-mine the mountain and reduce it by twenty-five hundred tons every hour for twenty to forty years. Seven shoreline crushers would grind the rock to gravel, which would be shipped abroad and dispersed throughout concrete foundations, roadbeds, and asphalt toppings. The project sparked resistance, and it revived modes of identity among some of the Indigenous Mi'kmaq, to whom it meant much more. Kluskap's Mountain was a center of cosmological significance as the home and place of return of the hero figure Kluskap.[5] The directors of Kelly Rock tried to negotiate their way to their desired outcome, but they gave up in the face of Mi'kmaq intransigence. They didn't understand what the issue was; they couldn't see how such a bounty of gravelly potential could be anything else. But mountains are many things, and things unto themselves. Following ecologist Aldo Leopold's guidance, if we "think like a mountain," we find connections between rocks, grass, deer, and wolves, between people and histories, and, in the words of a local journalist, a place where "every slug and shrew on the mountain has to be counted."[6] It is these life-sustaining connections that late modernity, in its blind optimism and technological hubris, throws aside as it lays an asphalt path ahead of its march toward the precipice.

Though we are fast approaching the end, we need to recognize that this has been a long march. The words "late" and "modern" camouflage the deep historical roots of this great unmaking and the processual nature of this phenomenon. Margaret Thatcher's statement that "there is no such thing as society" was as much a declaration of victory as it was a declaration of war on the connectivities that sustain and are sustained by humans and other organisms. Thatcher's words heralded a frightening acceleration in the process of dismantling the world, but the process itself reaches as far back as our ability to perceive connection allows us to see. Tony Judt sees the emergent individualism of the 1960s as integral.[7] The profound generational gap of that time was in no small part due to a different perspective on responsibility. Whereas the postwar era had focused the concern of the populace on the interests of the collective, the 1960s brought a radical shift toward individual freedoms. But these are merely salvos in an ontological barrage that has been fired upon humanity for centuries. René Descartes is certainly one hero of the unmaking. He sought to separate his humanity from earthly life, his ideas from history, and, in shutting himself up in a dark room, his very thinking from his corporeality. He set the ego apart from all else and reduced other animals to machines that showed motility only in response to the knives of anatomists. Historian Donald Worster locates unmaking's genesis slightly earlier, in the fifteenth century, as capitalist modes of production replaced mixed farming.[8] Over the six hundred years since, the complexity of the barnyard has given way to vast acreages of monoculture crops, where farmers are industrialists who don't even eat what they produce. Cognitive historian Jeremy Lent sees its beginnings much earlier, in the proto-Indo-European expansion and the dissemination of dualisms such as good/evil and mind/body.[9] According to Lent, late modernity is directly descended from the model of a split cosmos that the Greek and Vedic traditions inherited.

But I think unmaking stretches back even further, to the beginning of the Neolithic some ten thousand years ago, when a man and his neighbor first partitioned their land, as though birds, pollen, and threads of fungi didn't exist.[10] It's an ancient process, but one that only in the crisis of the Anthropocene is finally realizing its destructive potential.

So what exactly is unmaking? It is *the reduction of things to manageable units, a disregard for the threads that connect those things to their universe, and a shifting of those things into abstraction.* It isn't necessarily a physical act—it can be done conceptually—but it has consequences in the world of things. Debbie Rose talks of fragmentation, discontinuity, and interchangeability.[11] She points to the concept of the level playing field and its rendering of participants into interchangeable, substitutable units, as though familial and community ties, cultures, and histories are illusions. In her eyes, the level playing field is "one of the most pervasively deceitful of the great lies of this era."[12] Yet we are so inured to the deceit and so deeply entwined in its mesh that we cannot see the lie. Unmaking is reiterated and reinforced in our lives so pervasively that we come to see it as an objective reality, as if this is the only way the world can be. Take DNA. The ways in which genes interact with toxins, enzymes, viruses, and other genes during development are incredibly complex. Yet we are sold a line that our DNA is who we are. I can register with an ancestry website and order a DNA test kit. I spit in the sample tube, stick on a barcode, and send it back to have my DNA analyzed and compared with other samples taken from across the world. The ancestry folks acknowledge that human variation is only evident in 0.1 percent of DNA, but the results are presented so that a small portion of this 0.1 percent takes up 100 percent of a pie chart. By means of this chart, my entire self is abstracted, disembedded, and separated into discrete pieces of the pie: 70 percent here, 25 percent there, 5 percent there. What it implies is that this is where my ancestry lies and that, by extension, this explains who I am, but this isn't the case. The result simply shows where some of my contemporary genetic relatives (at least those who have been added to the database) live. Yet these relations are diffused into the abstractions of discrete geographical units that become concrete only for someone coloring in a map. My DNA has been disembedded from its context, my life history, my environment, and the food I eat and reduced to a set of data points among others that only exist in abstraction.[13] The test is sold as a way of connecting me with relatives I never knew existed, but in reality it reiterates unmaking, fostering disconnections with its implicit message that my DNA defines who I am, as though genes exist in isolation from everything else.

Unmaking permeates our education systems. As a child progresses through the school system, she is increasingly indoctrinated in the dogma of individualism. Scholarly excellence is measured as individual excellence, and the entire system of testing and grading is aimed at fragmentation. Rather than nurturing initiates, the education system pits individuals against one another and even against past students. This paradigm is so pervasive that we can't help but think that it's the only show in town. But it is in fact an ideology that must overcome human nature itself. To be human is to be collaborative, and this goes back to our very origins. Our ancestors didn't adapt to the east African savannah by taking turns wrestling lions. Cooperation and collaboration define humans even in comparison to other primates.[14] Yet contemporary scholarship treats this collaborative aspect of humanity as a novelty that can be tacked on to a curriculum. Even where students engage in joint projects, they are graded individually on the outcomes of the projects rather than on the degree to which they were collaborative. We emerge from the school system unmade and isolated, with a worldview that nucleates us and blinds us to connectivities, with our ethics neutralized, our histories erased, and our selves primed for reduction into something measurable and comparable. Job ready.

But if unmaking is not a natural human condition, how is it that people so readily adapt to the paradigm? Jeremey Lent maps the fusion of biology and history that brought us to this late modern cognitive impairment. According to Lent, the late Pleistocene saw humanity cross a cognitive Rubicon in terms of making meaning of the world. Building on the innate capacity of organisms to create heuristics for their interactions with their environments, humans took what he calls the "patterning instinct" to the next level by evolving feedback loops by which the instinct could be both co-opted and reinforced by cultural learning.[15] The key is in the prefrontal cortex, the overgrown part of the human brain that mediates and merges inputs such as vision, hearing, memories, emotions, passions, and urges. The prefrontal cortex is where these assorted inputs are made sense of and combined into a unified narrative; it is the part of us that makes sense of the world. Crucially, for us humans, we do this both instinctually and at a level of complexity that makes for surprising, complex, and potentially dangerous outcomes. As infants, our patterning instinct seeks out rules and consistencies in the environment while it discards excessive noise that contradicts the prevalent inputs. We absorb grammatical rules particular to our mother tongues and incorporate these not just into our language but into our worldviews. Lent's example is the definite article in ancient Greek, which facilitates the abstractions of Greek philosophy. We take the ideas of Greek philosophy for granted, but they could only emerge in a culture with a particular cultural-linguistic framework. Lent traces the Western cultural framework from ancient Greece, on through the biblical era, medieval Europe, and the Enlightenment to our present cognitive predicament. He shows us how the Western patterning instinct has attached itself to a dualistic worldview that holds rationality separate from materiality, mind from body, and man from nature. In this, Christianity and science, while seemingly at odds, are historical bedfellows. They both derive from Plato's idea that there is a rational realm from which the material can be made sense of.

Lent is clear that science is not itself unmaking. While it rides on the coattails of the Enlightenment, science is merely a tool in the process. Anybody who blames science for the fragmented world we live in is cursing the hammer for hitting their fingers. On the one hand, science gives us a means by which to reduce the world, but, on the other, it gives us a means by which to realize its complexity. Science provides the tool kit that reveals the multiple nodes of connected networks and the remarkable relatedness among nodes that constitute systems. In the right hands, science shows us connectedness within the deep complexity of food webs. It shows us how there are rarely more than two degrees of separation between species in terms of what they eat. In other words, if a particular species is not eating or being eaten by another species, it is almost certain that it is eating a species that the other species eats, being eaten by a species that the other species eats, eating a species that eats the other species, or being eaten by something that is eaten by the other species.[16] Ecologies have Kevin Bacon at every second step. And these are just food webs. There are countless ways in which species rely on or cooperate with other species for shelter, protection, and reproduction. What's more, the degrees of connectedness are more likely to be fewer where there is higher biodiversity. Science shows us how extinctions never happen in isolation, how they ripple through ecosystems, affecting everything. It shows us how food webs interact with climate, which interacts with people's use of air conditioners. When you burn fuel to cool your house, you are contributing to climatic shifts that are sending boreal fish poleward, where they have an impact on other species in arctic food webs.[17] Science has the resources to reveal complexity and show us a world of connections. And unlike unmaking, which is an ideology, science is neutral.

Unmaking's latest and potentially most destructive manifestation is neoliberalism. Never heard of it? Neoliberalism is an ideology born of a meeting of economists, philosophers, historians, and businesspeople at a resort in Mont Pelerin, Switzerland, in 1947. The meeting, organized by economist Friedrich Hayek, constituted the founding of a group that came to be known as the Mont Pelerin Society. The experience of the Second World War had raised concerns about the extent to which governments, no matter how well intentioned, were capable of impinging on individual freedoms. Hence Hayek's intention was that the Mont Pelerin Society would find ways in which liberalism might enrich the wider society by empowering individuals and limiting government intervention, how security might be reconciled with freedom to the benefit of all. Funded by wealthy benefactors, the supposedly apolitical society convened biennially to nut out the ways in which its vision might be transformed into public policy. By the 1980s, the ideas of the Mont Pelerin Society had crystallized, and world leaders were buying into the neoliberal model. The governments of Thatcher and Reagan actively pursued privatization of public services and utilities, deregulation of industry, reduced public spending, reduced taxes, individualization of the labor market, and the free movement of money and goods (but not people) across national boundaries. Neoliberalism provided an antidote to communism. The pursuit of private goods over public goods became inherently good, and the fall of the Berlin Wall in 1989 signaled a victory for the neoliberal ideal, at least in the eyes of its ideologues. But ideas and practice are never truly aligned. Among the critics of neoliberalism, George Monbiot shows that individuals have not been freed but rather have had their bonds shifted from states to corporations.[18] Workers are under the constant scrutiny of performance reviews and assessments, surveillance, audits, productivity measures, and monitoring that make communist bureaucracies look like fumbling amateurs. Neoliberalism not only celebrates selfishness but takes it as a given and undermines any notion of trust in workplace relations. It isolates people and pits individuals against one another, constantly raising the stakes by maintaining the necessary purgatory that is unemployment, with its own levels of monitoring. Neoliberalism has atomized people, increasing anxiety, social phobia, and depression while decreasing equality and social mobility. It concentrates power not in government for the people but in the hands of a few wealthy individuals.[19] When people fail in this system, they become social pariahs. The inequalities created and exacerbated by neoliberalism are more than disparities in wealth—they are a statement of disconnectedness, a denial of community and of shared responsibility. At the local level, responsibility is delegated to the individual, while power is taken away. Simultaneously, international institutions and actors divest themselves of responsibility and acquire more power.[20] The doctrine of personal responsibility frees governments from liability for their failures and leaves social welfare to charity organizations, effectively shifting taxation from the entire community to only those in the community who care. Charities tax kind people. Meanwhile, the selfish co-opt the system to turn money into an abstraction and move it around the world so it won't have to be shared with those on whose backs it is generated. Anna Tsing speaks of globalization in terms of flows. Things are remade as they globalize, but these flows are necessarily directed away from somewhere and from someone.[21] In forging global connections, they are always at the same time modes of unmaking. At the New York Stock Exchange, men in suits ring a bell and cheer the breaking up of the world into tiny bits that are exchanged for money and shuffled around like cups in a shell game. Restrictions are lifted from international actors, while others are more heavily imposed on powerless local actors in the form of impediments to transnational movement and collective action. The ultimate irony is that the reaction to neoliberalism has been a return to the populist, nationalist ideologies that led to the crisis for which neoliberalism was supposed to be an antidote.

For those who see the negative consequences of neoliberalism as a necessary price of personal freedoms, there is a flaw in the doctrine, and it has broader implications that will affect everyone, no matter how transnational, no matter how effective their defensive walls and security cameras—a flaw that emanates from its humanist underpinnings. Neoliberalism has created a technological juggernaut that, apart from enslaving its own consumers, threatens to destroy the earth systems that struggle to sustain its unfettered, intrinsically necessary growth. Even where it is heavily regulated, industrialization comes at a heavy cost to ecologies. Where regulations are lifted and freedom is underscored by noninterventionist governments, the environmental costs are multiplied. Unfettered corporations are free to strip landscapes of their resources because, for the time being, there are plenty more landscapes to be had. And they do so without regard for the complexity of earth systems that would interfere with the bottom line. This is all because the neoliberal doctrine is underpinned by a deep-rooted Euro-American worldview that reflects its membership and its geographic centers of power: the worldview that humans and nature are separate things, that profit can be separated from ecology, and that stones can be separated from mountains.[22] Val Plumwood refers to the "ecological remoteness" that these global flows foster, where the ecological consequences are felt at neither the proverbial nor the literal coal face.[23] Governments, producers, and consumers are invariably at a distance from the effects of their involvement in the neoliberal game, which lends an ecological irrationality to their decisions. Like the schizoid man who cuts off a finger that he thinks is not his own, neoliberalism—in its multiplicity of separations—reconfigures, abstracts, and demolishes the world on which its existence depends.

In Australia, neoliberalism was manifested in government policy beginning in the mid-1980s. The key difference between Australia and the United Kingdom and United States is that these changes occurred under a Labour government with initial support from the trade union movement.[24] The Labour government floated the Australian dollar, deregulated industry and the banking sector, abolished free tertiary education, dismantled tariff protections, targeted welfare, reduced taxation, and restricted spending. Labour corporatized government departments and set in motion the privatization of public assets. It dismantled central wage indexing and introduced "enterprise bargaining," a means by which workers could be disengaged from union influence and compelled to negotiate directly with employers.[25] During this time, the federal government formed the Rural Industries Research and Development Corporation (RIRDC).[26] This government body was tasked with facilitating and sponsoring research to promote a "profitable, dynamic and sustainable rural sector."[27] As a locus of research on new farming methods, the RIRDC was integral to the growth of canola farming in Australia, along with increased rice yields and improvements in production of chickens through the development of vaccines. Outwardly, the RIRDC might seem to have little to align it with neoliberalism. A government-run body offering money for research projects is not a shining example of the neoliberal model. But seen through a lens of unmaking, they align quite neatly. The RIRDC subscribed to a view of nature as raw material, the components of which could be taken up and manipulated by human hands for economic benefit. It saw singular plants and animal species as separable, farmable, and scalable. Under the corporation's auspices, the entire Australian ecology was opened up to analysis and exploitation, and Australians were opened up to the possibilities this corporation offered. Since federation, Australians had increasingly taken pride in native animals. By the 1990s, nativeness was seen not just as intrinsically good but as somewhat compatible with the colonial project. Native animals became promising for farming in Australia because of a perceived compatibility with the Australian environment. As a result, deliberate, calculated attempts were made to domesticate at least a few species of Australia's fauna.[28]

I will return to the Australian domestication project in due course, but first I want to frame domestication writ large as a product and producer of unmaking. In terms of turning points in human history, the domestication of plants and animals is one of the most profound and potentially the most disastrous. Jared Diamond refers to the transition to farming and pastoralism some ten thousand years ago as "the worst mistake in the history of the human race." For comparisons, he points to modern-day hunter-gatherers and demonstrates that even though they live in extremely marginal habitats, they have rich and varied diets compared to the diets of agriculturalists. What's more, they expend comparatively little effort to enjoy these diets. These people spend no more than twenty hours per week getting food. Tim Ingold argues, further, that this is not even considered time spent working.[29] The other basis for Diamond's claim is the archaeological record, specifically the bones and teeth of ancient humans. When you compare the remains of ancient hunter-gatherers with those of ancient farmers, you find some telling differences. In prehistoric Greece and Turkey, for example, there was a crash in average height after the adoption of agriculture. While hunter-gatherers averaged 175 and 165 centimeters (69 and 65 inches) in men and women, respectively, these averages dropped by as much as twelve centimeters (nearly five inches) after the population adopted agriculture. In fact, even modern-day Greeks remain below the average hunter-gatherer height. Another change shows up in dental and bone pathologies. Researchers have found that when hunter-gatherers in Illinois adopted intensive maize production, the emergent agriculturalists experienced a 50 percent increase in enamel defects, a 400 percent increase in iron deficiency, and a 300 percent increase in bone lesions and degenerative conditions indicative of labor in the fields. After these people adopted agriculture, their average life expectancy decreased by six years. In terms of health, transitioning to farming has high costs. The risk of starvation rises, pathologies associated with hard work multiply, diet becomes less varied, and close contact with animals, in combination with higher population densities and more complex trade networks, increases the spread of infectious diseases. Measles, pertussis, smallpox, tuberculosis, taenid worms, and falciparal malaria are all products of domestication and the societal changes that resulted.[30]

Anthropologist Clark Larsen says that "although agriculture provided the economic basis for the rise of states and development of civilizations, the change in diet and acquisition of food resulted in a decline in quality of life for most human populations in the last 10,000 years."[31] This statement is on point as far as human health goes, but it is odd to contrast states and civilizations (implied as a good) with poorer quality of life (clearly not good), because the latter is typically a consequence of the former. Civilizations certainly conferred benefits on the wealthy, and more recently on the middle classes, but not without a cost. They also facilitated large-scale warfare, profound global inequality, widespread poverty, oppression of women, slavery, industrialized animal cruelty, and environmental destruction, all of which were in turn made possible through the domestication of plants and animals. Prior to the agricultural revolution, there was not a lot of incentive to accumulate goods. Mobility is crucial in exploiting wild plant and animal resources, and lugging a sackful of grain around would decrease hunter-gatherer mobility exponentially. But once human populations were able to take animals and plants out of their ecological contexts and bring them closer to home, they instigated profound changes in how societies were organized.[32] Domestication was a double-edged sickle: on one side, it conferred control of resources and accumulation of wealth on particular individuals; on the other side, it wrested self-sufficiency from people and made them dependent on the individuals who controlled the resources. No longer could a person simply walk away from asymmetrical power relations and find her own subsistence. Domestication fostered new divisions of labor, making the vulnerable dependent on the powerful and tying them into society by inhibiting their self-sufficiency.[33] This was not an inevitable consequence of domestication, and there are plenty of examples of agrarian societies in which overaccumulation by particular individuals is limited. But where there were no limits, power became cumulative. In places like Egypt, East Asia, the Middle East, and Mesoamerica, control of resources led to massive social inequality and the concentration of wealth in a handful of elites. Such power can only be maintained by force, but fortunately for these elites, domestication fostered the capacity to maintain standing armies. These armies became tendrils of expanding states, used to wage war against neighbors and accumulate land, resources, and slaves. We might look at our modern society and appreciate that civilization has conferred a better quality of life for at least a portion of the human population, but it has come at the cost of an awful lot of warfare, famine, disease, inequality, and oppression, none of which has gone away.

Plant and animal domestication not only wrought changes to the ways that societies were organized and personal liberty was understood; it also changed human bodies. Anthropologist Helen Leach shows how, in addition to a reduction in stature, human populations in domesticated settings exhibit other morphological changes that parallel the changes in their animals.[34] One such trend is increasing gracilization, in which the robusticity of the hunter-gatherer skeletons gave way to more gracile forms. Across these populations we find reduced epiphyseal bone diameters, smaller brow ridges and mastoid processes, smoother bone surfaces, and smaller brains. Post-domestic humans also exhibit craniofacial shortening, tooth crowding, reduction in tooth size, and in some cases fewer molars. Others have found changes at the molecular level. For example, people of the Neolithic who raised cattle and sheep and drank milk after weaning were under selection for lactose tolerance beyond childhood. Among modern humans with pastoralist ancestry, there is a much higher percentage of adults exhibiting an enzyme that helps break down lactose in milk. This adaptation to drinking milk into adulthood developed at least once in Europe and four times in sub-Saharan Africa. If you can drink milk without running for the toilet, chances are your ancestors herded cattle (or acquired milk from people who did). Another change common to those with agriculturalist ancestry is in the number of copies of the gene AMY1, a gene associated with the enzyme amylase, which helps with starch hydrolysis. In agricultural populations with high starch diets, there is a higher frequency of people with an increase in copies of this gene.[35] Changes also resulted from the infectious diseases that domestication fostered. Among Europeans, there is a high frequency of allele deletion of CCR5-Δ32. The high frequency of this deletion, which has been found to confer resistance to HIV-1, probably arose in response to extreme selective pressure brought about by smallpox.[36]

As for domesticated plants and animals, the changes wrought to their bodies are a matter of public record.[37] Maize, for instance, is a bloated mutant compared to its wild cousin teosinte. It has become so dependent on cultivation that it cannot reproduce in the wild. Other cereal grains, such as wheat, are similarly bloated with carbs and require threshing to separate them from their stalks. Most domesticated animals exhibit profound physical changes in comparison to their ancestors. Domestic cattle are far smaller than their ancestors the aurochs, who were huge creatures. A bull aurochs stood at 200–220 centimeters (79 to 87 inches) at the shoulder.[38] Horn size and shape have also been radically altered in cattle. In sheep, size is also reduced in comparison to mouflons, and specific morphologies have been altered. Some sheep carry massive amounts of fat in their tails, while others have folds of skin for carrying copious quantities of wool. Horses were initially uniformly small, but domestication has seen a plethora of shapes and sizes emerge, from miniature ponies to massive draft horses. As for dogs, although there are now breeds that outweigh wolves, the earliest dogs are distinguished archaeologically from wolves by their smaller size, shorter snouts, and crowded dentition. And some modern dogs are almost unrecognizable as wolves, despite their being the same species in molecular terms. It seems absurd, but a toy poodle is genetically a wolf.

The evolution of dogs from wolves in fact marks the beginning of the story of animal domestication. Almost everybody agrees that wolves were the first animals to be domesticated. The only dissenters are those who say it wasn't wolves but dogs who became domesticated. Raymond and Lorna Coppinger argue that wolves initially evolved into dogs by hovering around settled human communities. Then, after having evolved into dogs, these canids became domesticated. The problem with this argument is that it assumes that there were settled human communities producing large enough quantities of waste for the wolves to scavenge. It might hold up if domestication of dogs occurred in agricultural communities, but it's harder to prove with hunter-gatherers.[39] Hence the timing and location of dog domestication becomes an issue, and an unresolved one at that. Skeletal evidence has proto-dogs appearing in the archaeological record in Europe as early as thirty-one thousand years ago, well before people became settled agriculturalists.[40] Meanwhile, molecular evidence suggests both earlier and later dates. Carles Vilà and colleagues sequenced mitochondrial DNA of canids and, using assumptions about mutation rates, estimated that domestication occurred in multiple locations as early as 135,000 years ago. Jun-Fen Pang and colleagues rejected that estimate because of limited phylogenetic resolution; based on their study of homogeneity among dog and wolf haplogroups, they argue that the domestication of dogs occurred in southeastern China between sixteen thousand and five and a half thousand years ago. They also note that the founder population was genetically diverse, suggesting that there was some cultural trait that saw repeated domestication events. They also suggest that dogs might initially have been domesticated for their meat. Raymond Pierotti and Brandy Fogg argue for an earlier time line and a different process, one that involved mutualism rather than consumption. Presenting examples of wolves cooperating with ravens and African wild dogs cooperating with jackals, they argue that humans and canids are incurable cooperators who couldn't help but form mutually beneficial hunting strategies. These strategies in turn brought wolves closer and closer to humans, and the more sociable among them became the ancestors of dogs.[41] I return to this point in the next chapter.

It was not until after the agricultural revolution some ten thousand years ago that other animals became domesticated. At least this is the period in the archaeological record where the morphological and demographic changes associated with domestication begin to appear.[42] The first of these were goats and sheep who evolved out of bezoar goats and Asian mouflons, respectively, in the mountains bordering the Fertile Crescent—the arc of land extending from southeastern Turkey to the Persian Gulf—which saw the emergence of cereal agriculture. Melinda Zeder and others suggest that these animals followed a different domestication pathway from that of dogs; Zeder calls it the "prey pathway."[43] In this scenario, humans developed hunting strategies aimed at maximizing their take of goats and mouflons, and these strategies evolved into forms of herd management. For example, hunters might have driven herds into ravines or corrals and kept them contained while they harvested animals over a period of time. It wouldn't take long for those hunters to realize that if they hung on to a few individuals and let them breed, then they wouldn't need to go chasing herds around the mountains anymore. Cattle and pigs were later domesticated in the same region.[44] Interestingly, it looks as though cattle were independently domesticated in southern and eastern Asia as well. Once the gates to domestication had opened, other species flowed in. Horses were domesticated in multiple locations across the Eurasian Steppe after cattle and sheep had been introduced.[45] Asses were domesticated in northern Africa and, with camelids, in the Middle East, rabbits and ducks in Europe, geese and guinea fowl in Africa, chickens in Southeast Asia, carp and silkworms in East Asia, pigeons and bees across Europe and Africa, and, eventually, reindeer in the Arctic north.[46] In the Americas, llamas and alpacas were domesticated around five thousand years ago, followed by turkeys, guinea pigs, and Muscovy ducks. Others who could be added to the list are elephants, fallow deer, cochineal beetles, snails, and swans, although some would argue that these species are only semidomesticated.

The list of animals domesticated prior to the nineteenth century is not particularly long considering the ten-thousand-year time span, but it grew dramatically in the past two hundred years. Since the Industrial Revolution, more than one hundred animal species have been added to what

Melinda Zeder calls the "domestication superhighway," by which she means the intensification of domestication processes under the specific and deliberate direction of human communities.[47] The vast majority of these, mostly birds, are in the pet trade. Other species are farmed for meat, and given the intensive conditions of their farming, could be considered to be under domestication processes. Many of these are fish species such as basa, salmon, barramundi, perch, grouper, bass, mullet, and snapper, all spearheading the movement of farming into the world's oceans and waterways. Then there are animals such as rats, mice, hamsters, and chinchillas that are being domesticated for lab research. Several other species provide multiple products: emus, ostriches, and rheas provide oil, meat, eggs, skins, and feathers; quails provide meat, eggs, and feathers; crocodiles provide meat, teeth, and skins; deer provide meat, skins, and antlers. Some species are being domesticated for their services. Birds such as the red-legged seriema and the evocatively named southern screamer are apparently used in South America to guard against foxes. This is the new wave of domestication. Not only are the novel species and their human overseers pushing the parameters of what sorts of animals might become domesticated; they are challenging traditional definitions of domestication.

In the course of my work, I encounter a lot of different ideas about what exactly domestication is. I find that a lot of people think of it in terms of "keeping at home," in the same way that women in the first half of the twentieth century were characterized as domesticated. So when I speak of Australian native animals being domesticated, some people assume that I mean something like spotted-tail quolls being kept as replacements for pet housecats. But the sort of domestication that I have in mind is broader in scope and much more difficult to define. In the edited collection *The Walking Larder*, Juliet Clutton-Brock famously defines a domesticated animal as "one that has been bred in captivity for purposes of economic profit to a human community that maintains complete mastery over its breeding, organization of territory, and food supply."[48] This is a difficult line to hold. As Agustín Fuentes says, monkeys bred for research purposes fit this operational definition of domestication, but few would consider them domesticated.[49] The case of farmed emus and crocodiles is similar, as I will show. At the same time, some animals who most people would accept are domesticated fall outside this definition. Take the horses among the Oromo people of Ethiopia. They are not usually bought and sold, and given the labor and food invested in them they do not provide much in the way of "economic profit."

They confer prestige on, and share identity with, their owners, but they are not kept for making money.[50] Certain horses are selected from the herd for particular physical characteristics like height and color, but their breeding isn't controlled. They are free to breed with whomever they chose. No fences pen these horses in. There are fences to keep them out of cultivated fields, but as with their breeding choices, Oromo horses have considerable freedom. They choose much of what they eat. Their human masters feed them special treats like barley, but otherwise the horses graze on whatever is edible in the fields. Given that every one of Clutton-Brock's criteria for domestication is missing, can we say that these horses are not domesticated? Could they be considered wild horses held captive? That's another difficult call to make, in that there are no wild horses in Ethiopia—just a single feral population of a couple dozen horses at Mount Kundudo. In fact, horses were never native to Ethiopia—they were introduced to the country by humans sitting on their backs. I suggest that rather than try to reconcile these horses with Clutton-Brock's definition of domestication, we should reconsider the definition.

Pierre Ducos gives us a somewhat looser definition, arguing that a domesticated animal is one that is an object of human ownership. But I can think of a lot of animals who are owned by humans but wouldn't be considered domesticated—deer in aristocrats' estates, insects held captive in the glass jars of children, and the entire native fauna of Australia, owned by the Crown, are but a few examples.[51] Ownership of animals is certainly a corollary of domestication, but to say that it defines domestication is to put the cart before the horse. A definition needs to take into account the population levels at which domestication occurs and the nuances of human relationships with their domesticates. More recently, Melinda Zeder has defined domestication as "a sustained, multigenerational, mutualistic relationship in which humans assume some significant level of control over the reproduction and care of a plant/animal in order to secure a more predictable supply of a resource of interest and by which the plant/animal is able to increase its reproductive success over individuals not participating in this relationship, thereby enhancing the fitness of both humans and target domesticates."[52] This is a carefully crafted definition that borders on political correctness. "Some significant level of control" certainly encompasses more than "complete mastery," and a "more predictable supply of a resource of interest" is more nuanced than "economic profit," but there's something about "enhancing the fitness" that seems circular. Fitness in what context? If Zeder means fitness for the domestic sphere, then there stands the circularity: that's not enhancing fitness but altering it to fit within domesticity. We cannot say that domesticated broiler chickens, who grow so rapidly that they can't stand up, have had their fitness enhanced in any way other than as producers of flesh. I doubt that a broiler chicken would fare well in the forests of southern Asia or even in the yards of rural communities.

Some scholars are not even satisfied with the use of the term "domestication." Natasha Fijn proposes that it be dropped. She prefers the term "co-domestic relationship," which she applies to Mongolian herders and their livestock, because it highlights the mutuality of domestication. Fijn emphasizes the social dimensions of domestication, by which "mutual cross-species interaction and social engagement" foster relationships that are about more than reproductive success and fitness in humans and target species. In the course of her ethnographic fieldwork in Mongolia, she encountered a "we feed them and they feed us" perspective that was very much contrary to the Western view of domestication as based in human control.[53] As with Oromo herders, Mongolians give their livestock animals a great deal of autonomy in return for protection from predators, in this case wolves. The key aspect of domestication, then, is these animals' integration with human communities. Following a similar line of reasoning, archaeologist David Orton suggests that a definition of domestication should take into account the roles that domesticated animals play in structuring human societies. Rather than considering them objects of ownership, he presents a view of domesticated animals as "sentient property," as social actors in their own right, mutually shaping processes of domestication with their human caretakers.[54] With specific reference to the Vinča culture of the Balkan Neolithic period, Orton argues that the key difference between domesticates and their wild counterparts is the space for direct intersubjective relations. This definition emphasizes mutuality in terms of social relations between humans and domesticated animals—it allows animals to shape the conditions of their own transformation. What I find wanting, though, in all of these definitions of domestication is an account of the negative space created by these relationships between humans and domesticates. I suggest that it is that which is taken away in the course of enfolding animals into human societies that in fact defines domestication.

In the 1950s, a Russian geneticist sought to bring together the social and biological aspects of domestication in a unique experiment that was to last more than fifty years. The geneticist was Dmitri Belyaev, and the experiment came to be known as the farm-fox experiment. Belyaev was interested in unanswered questions about the initial process of domestication. First and foremost: how did the process get started? How did ostensibly wild animals lose their fear of humans and attach themselves to human communities? And why had only a few species been domesticated over a period of ten thousand years?[55] What was it that facilitated domestication in these species and not others? There was also the question of traits that appeared in domesticated animals that cut across species boundaries. Floppy ears, blotchy coloration, curly tails, and babyish faces are common to many domesticates. Were all of these traits deliberately selected for by farmers over the history of domestication, or are they corollaries of something else? Another question was why many domesticates were able to breed more than once per year and outside the breeding seasons of their wild cousins. This question had a dual role, in that it served to mask the intentions of Belyaev's experiment. In the Soviet Union at the time, a well-connected plant breeder named Trofim Lysenko had found favor with Stalin and had risen to a position of power in the Soviet scientific community. He was strongly opposed to geneticists because they threatened to disprove his claims about plant breeding, so Belyaev needed a cover for his experiment. The official line he gave was that he was running an experiment to increase the production of fox fur by seeking to discover the mechanisms by which foxes might breed more than once per year. And he had a hunch. He guessed that the answer to all of the questions lay in the defining characteristic of domesticated animals: tameness. He thought that this trait must have been crucial in the history of domestication, because animals who lacked it would have been intractable and unmanageable and therefore selected against. So he designed his experiment to simulate what he thought was key to the process of domestication. He had noticed that among farmed foxes and mink there was a gradient in terms of tameness, with extreme aggressiveness at one end and extreme fearfulness at the other. Somewhere in between were animals who were calm around people, and this variation suggested that it would be possible to select for tameness. He asked the breeding team at the silver fox farm at Kohila in Estonia to choose from among their foxes those who were calmest in the presence of humans and then mate these foxes with one another. The offspring of these matings were to be similarly assessed and again bred with one another (with some introgression from other farm foxes selected for tameness so as to counter problems of inbreeding).

The method for testing tameness in the foxes was simple. Farm workers were instructed to approach each fox slowly and reach inside the cage with food held in a gloved hand. Most foxes shied away, attacked the hand, or retreated snarling, but a few were neither highly aggressive nor reactive.[56] These were selected for breeding. It took only three breeding seasons before the experimenters began seeing foxes who were tamer than their parents. But Belyaev wanted to scale up the experiment. He moved it to Siberia, where animal behaviorist Lyudmila Trut would oversee the selection of foxes on a commercial fox farm. By 1961, the experimental population of foxes was one hundred females and thirty males. Not only was each generation of tame-selected foxes showing less fear and aggression toward humans, but the farmworkers were petting them more. The foxes' tameness was eliciting nurture from humans, creating a feedback loop. But it wasn't just behavior that was changing from generation to generation. Belyaev and Trut found that some of the female foxes were mating a few days earlier in the season than before, and producing on average more pups per litter. There were also behavioral and morphological changes that hinted that Belyaev was on the right track. After the fourth generation of pups was born on the Siberian research farm, Trut noticed that one of the pups, named Ember, wagged his tail vigorously when she approached him. This is not normal behavior for foxes. After the sixth generation, some pups were rolling onto their backs inviting tummy rubs, and whining when their human caretakers walked away. By the eighth generation, some of the wagging tails were also curly. By this time, Belyaev hypothesized that the key to domestication was the timing of certain traits. The ways in which pup-like behaviors persisted and mating came earlier suggested that hormonal changes were occurring. What's more, hormones are associated with stress, which the tame-selected foxes were coping with better than their aggressive or fearful cousins. Hormones are regulated by genes, so selecting for tameness seemed to confirm that it was not mutations that characterized domesticated animals but the timing by which genes regulated hormones. When they compared levels of stress hormones between tame-selected and control foxes, the researchers found that these hormone levels spiked much later in the tame foxes and at only half the level of the controls.[57] Belyaev coined the term "destabilizing selection" to account for the way in which selection for tameness was not fostering genetic change in classical terms but was altering the timing of gene expression, which in turn was affecting the timing and intensity of hormonal expressions.[58] When the tenth generation of pups was born, this time in a dedicated research facility in Akademgorodok, a Soviet model science city, one of the pups exhibited floppy ears well beyond the time when fox pups' ears have usually straightened out. Another had piebald coloring. By the fifteenth generation, Trut had taken one of the pups home as a pet. In less than two decades, the farm-fox experiment had created what was, for all intents and purposes, a dog.

Destabilizing selection by definition implies that, outside human-controlled selection for tameness, there is a stabilizing force that selects against the sorts of traits that emerged in the farm-fox experiment. We might call that force the environment. Climate, predators, prey, parasites, fox population densities, and intraspecies competition combine in the habitats of silver foxes to set very narrow parameters as to when genes are allowed to switch on and off and as to when particular traits must be expressed or dissipated. In the Eurasian Steppe, a tail-wagging, approval-seeking, sycophantic fox would be at a disadvantage as an adult in comparison to a dour hunter who was wary of other species. A litter of pups born a couple of days too early would stand a greater chance of being frozen or starved to death. As Paul Shepard says, "Niches are hard taskmasters."[59] Fox ecologies make strong demands of their inhabitants, and given environmental conditions that don't change dramatically, any outliers are starved, brutalized by their siblings, snatched up by wolves or eagles, or refused a chance to mate by other foxes. In this respect, we need to understand that it was not just selection for tameness that created the floppy-eared, piebald, tail-wagging pups of the farm-fox experiment, but the separation of those fox populations from the ecologies of their wild cousins. Of course, the farm-fox experiment was aimed at understanding the genetic changes associated with domestication rather than at theorizing about domestication itself, but it highlights the fact that before selection for domestic traits can begin, the subject population must be separated from its ecological context.[60] Leach refers to this as "operational selection."[61] This is the removal of, or protection from, selective pressures that would otherwise select against traits such as docility, small body size, and delayed maturity. The farm foxes were kept in sheds with fenced exercise yards—no predators could penetrate those defenses. Pups abandoned by their mothers were nursed by humans and insulated from the cold. The foxes were given veterinary treatment when ill. They were separated from mate choice in that their matings were largely decided by humans. They were relieved of the need to hunt for food. Without these kinds of separations, the farm-fox experiment would have failed and selection for tameness would have been futile. Fox ecologies would have shut down the experiment like no corrupt bureaucrat ever could.

For animals such as farmed fish, it's unlikely that socialization to humans will ever occur, let alone that they will exhibit piebald coloring or curly tails. But something must be happening in the contexts in which these animals are associated with humans that is creating differences from a wild state and resulting in population changes. So rather than focus on what is produced as a result of enfolding animals into a human niche, it's more enlightening to focus on what is destroyed, and that is ecological complexity. This is where I see the key to domestication: *it is the separation of animals from their traditional ecologies and their incorporation into a human niche*. Such separations occur at the level of a reproductively viable population and adapt the animals to a novel, anthropogenic feeding, reproductive, and social paradigm—what Jean-Denis Vigne refers to as an "anthroposystem."[62] In most cases, the primary separation is from predators. This is the case with Oromo horses, Mongolian cattle, and just about every other species. In other cases, the separation is from a traditional prey base, as with carnivores like dogs and cats, and now also farmed fish, who are fed pellets. Separation in terms of feeding paradigms occurs even at the smallest level, where food is supplemented or animals are kept from what would otherwise be their natural feeding grounds. And it occurs on a continental scale when animals are separated from their traditional habitats and shifted to landscapes and climates different from those to which their ancestors were adapted. Inevitably, the process of domestication dumbs down these animals' ecologies. Rather than inhabiting a complex web of intricate relationships with other species, the animals are brought into simplified human ecologies. Certainly, domestication to a human niche involves establishing connections with novel organisms. As Agustín Fuentes and Eduardo Kohn note, humans have a particular knack for "altering, entangling and reconfiguring" ecologies in ways that profoundly alter the organisms with whom we share the domestic sphere.[63] In the barnyard, there are zoonoses, parasites, altered habitats, and novel feeding regimes that exert all manner of novel selective pressures on animals.[64] But these domestic niches are hack ecological networks held together with string and glue. They are highly unstable and dependent on human intervention rather than existing in a balanced state. Domestication, by separating animals from their traditional ecologies, produces ecologies of dependence in which human intervention constantly ratchets up the need for separation until the animals exist in ecological deserts. In this way, separation from ecologies forms the conceptual link between dogs, sheep, farmed fish, and pet parrots. They all fall under the rubric of domestication because they have all been separated from their traditional ecologies at a population level and inducted into a simplified ecological niche that is a product and producer of humans. As such, I would argue that these animals are being unmade.

In Australia, there is an expanding list of species that are being bred, farmed, and made into pets, and all of these are subject to separation from their traditional ecologies. Among Australia's mammal species, there are wallabies, gliders, and quolls. Among the birds are cockatoos, budgerigars, cockatiels, parrots, rosellas, doves, finches, quails, ravens, and ibis. Of reptiles there are bearded dragons, pythons, and blue-tongued skinks. Among the aquatic species are mulloway, rainbowfish, yellowfin tuna, grouper, barramundi, longfin eel, red claw crayfish, and freshwater crayfish. Even fruit flies are bred for laboratory experiments. This is a long list, and all of these species are worthy of investigation, but for reasons of limited resources and popular appeal, I restricted my study to a handful of somewhat charismatic species that are among the most likely suspects in terms of domestication—namely, dingoes, stingless bees, crocodiles, emus, and kangaroos. There is a long history of dingoes associating with human communities and being kept as pets after colonization, but recent times have seen increased interest in keeping dingoes as pets and breeding pure dingo bloodlines. Stingless bees have been domesticated in other parts of the New World, but only recently have they become popular in Australian farms and households. Crocodiles are being extensively farmed or ranched in Australia and overseas, initially in connection with conservation efforts but now as a commodity in their own right. Emus are also being farmed not just in Australia but worldwide, often in conditions markedly different from those experienced by wild emus. I added kangaroos to my list of subject species not because they are being domesticated but because they are not, although some people keep kangaroos as pets. The persistent wildness of kangaroos in the face of economic exploitation might shed some light on domestication processes in Australia.

I arrived at this topic through my involvement in the Human Distinctiveness Project at the University of Notre Dame, the brainchild of Celia Deane-Drummond and Agustín Fuentes. When I joined the project, they asked me to come up with a study on domestication broadly construed. My initial plan was to go back to Ethiopia and spend a year among the Oromo agro-pastoralists of Ethiopia, studying their complex relationships with their livestock animals and the hyenas who threaten them. I was particularly interested in what I might learn about early domestication as a process of excluding nonhuman predators from early domesticates' ecologies. But as I was planning the project, the security situation in Ethiopia rapidly deteriorated. Oromo people were staging mass protests against human rights abuses, government security forces were detaining, beating, and killing not just protesters but bystanders, and the place where I intended to do my fieldwork just happened to be ground zero for the resistance movement. So I had to rethink and relocate my project. It was then that I heard about people who kept native bees in Australia, and I realized that I had an opportunity to see domestication not after the fact but as a process in its early stages. In historical terms, domestication events are extremely rare—at least they had been until the nineteenth century—and I had a chance to bear witness to such an event as it was unfolding. I can't overstate the significance of this. As with the massive changes that are being wrought upon the earth by population growth, urbanization, and industrialization, the new wave of domestication promises to profoundly alter the ways in which humans engage with ecologies. We stand at the threshold of something akin to a Neolithic revolution; the way these processes unfold will resonate through human societies for millennia, should humanity last so long. The difference is that, unlike the first farmers of the Neolithic, we are in a position to grasp that there are potentially grave consequences to these actions.

In order to see these domestication processes through a critical lens, I needed some sort of comparison. Considering that Aboriginal people prior to colonization sustained ecologically sound relationships with all of the species examined in this book, I saw it as logical to compare modern domestication with traditional Aboriginal ways of conceiving of and engaging with the species considered here.[65] This is delicate work, because in terms of Aboriginal conceptions, it can be taken to imply that there is a singular, monolithic Aboriginal worldview and also that I can somehow represent it. I reject both of these implications outright. To suggest that a Yolngu woman in Arnhem Land holds the same worldview as that of a Koori man living in Sydney is both delusional and racist, and to suggest that I could put either of those views down on paper is defunct anthropology. Instead, what I seek to hold up as my critical lens is a hybrid perspective derived from what Aboriginal people have been trying to tell colonists for more than two hundred years. I'm not holding up this lens with any claim to its authenticity but rather using it to see more clearly Western ontologies that view Australia as a land that can be possessed and its animals as resources that can be domesticated. What matters here is to question colonial Australian ways of engaging with the land and to seek potential ways forward in terms of reconciling them with Australia's ecologies. At the same time, there's an indisputable difference between the ways in which Aboriginal people traditionally engaged with the land and the ways in which colonial Australians have done so. We have physical evidence of this difference in the massive ecological destruction wrought on the Australian continent since colonization. Deforestation, species extinctions, soil erosion and degradation, salinity, degraded river systems, industrial pollution, algal blooms, car parks—these are all recent phenomena that were unknown to precolonized Aboriginal people. Besides sanctioned genocide and continental land theft, colonization wrought ecological destruction on such a scale that were it done by a corporation today, the management would be headed for jail. So my project here involves seeking those aspects of traditional Aboriginal practices and ontologies that fostered and maintained healthy ecosystems. In pursuing this goal, I need to remain open to counterintuitive possibilities; I need to allow that what Western thinking might see as anathema to conservation might in fact make for healthy ecologies. For many Western conservationists, an ecosystem devoid of humans is utopic, yet within many traditional Aboriginal ontologies, humans are fundamental to healthy landscapes, not just as managers but as participants in ecologies.[66] For colonial Australians, introduced species are seen as inimical to conservation goals, whereas some Aboriginal people are able to incorporate these interlopers into their landscapes without any conceptual crisis, as long as the country remains healthy.[67] In this, Debbie Rose, with her concept of an Indigenous philosophical ecology, is again my guide.[68] She calls the Aboriginal perspective a metaphysics of pattern, benefit, and connection. In her work among the Yarralin people of the Victoria River region, Debbie articulates a nonhierarchical, nonglobal perspective on ecologies and the place of humans within them. She sees four areas of dialogue in which humans might be resituated in relation to landscapes and life systems: in the first, subjectivity exists not only in humans but among all participants in life processes; in the second, humans exist not at the apex of a food web but in complex, nonhierarchical patterns of localized connectedness among other organisms; the third concerns the concept of kinship and holds that morality cuts across species, operates laterally, and is mediated by connectedness; and the fourth calls into question human autonomy and examines the ways in which humans are called into action in relation to other beings. Rather than serving as a blueprint for otherness, an Indigenous philosophical ecology operates on the premise that humans are not exceptional but are integral to healthy ecosystems. In both Indigenous ecological and Western environmental perspectives, there is room for intensive human involvement, but while the latter sees humans as playing God as managers and system organizers, the former sees humans as participants, kin, integral members of ecological communities. Only in their physical manifestations are humans conceived of as different from other organisms.

While my sympathies lean toward Indigenous philosophical ecologies, I want to be clear that this is not a blanket idealization of Aboriginal society. Human societies are never ideal, and Aboriginal people were, and are, as flawed as any other. The ways in which humans incorporate cultural learning into the human niche makes for some pretty messed-up versions of self-organization. Ideas can take on a life of their own, and people can become subservient to cultural norms. Aboriginal people practiced and institutionalized violence not just in relations between groups but in sorting out internal disputes and meting out punishments.[69] As with many human groups, young girls were subject to oppression and forced marriage. And of course there were all of the petty rivalries and jealousies that characterize any human society. Yes, Aboriginal people have been subject to massive social, emotional, and physical displacement under colonization, much of which remains to be reconciled, but this does not mean that Aboriginal societies before 1788 were perfect.

I spent more than two years researching this book, though not in typical anthropological fashion. Normally, anthropologists embed themselves in particular places for long periods in order to learn the language and gain qualified insights into what the natives are about. I'm already a native among colonial Australians, so I was spared the disorientation, confusion, and intestinal parasites of that aspect of anthropology. Marianne Lien calls this kind of research "nearly home" ethnography.[70] I made trips from my home in northern New South Wales to places up and down eastern Australia, where I met with a range of people, from backyard beekeepers to government representatives.[71] I drank their tea and coffee, borrowed their time, and recorded their accounts of how they engaged with the Australian environment and its animals. In most cases I interviewed people only once, although there were a few I revisited. Where animals were being farmed or bred, I made several visits to give myself a chance to catch more detail. As a matter of anthropological procedure, I have given these people and their organizations pseudonyms unless they explicitly agreed to be identified. I've occasionally done the same for place-names. In every case I found these people helpful, passionate about what they do, and willing to share their stories. In this new wave of domestication, there are no villains, at least not in the dualistic sense. There are just people and animals, myself included, who are able to manipulate the process only insofar as their cultural frames of reference will allow. The new wave of domestication is not a carefully orchestrated program of change; it is a haphazard process that carries people, animals, and ecologies into unfamiliar landscapes. We can only hope that with a greater degree of understanding, those landscapes will survive the intervention.

2.

Dingoes

He was a beautiful creature—shapely, graceful, a little wolfish in some of his aspects, but with a most friendly eye and sociable disposition.
—MARK TWAIN, *Following the Equator*

Several thousand years ago, someone from Sulawesi landed a small seagoing craft on the northern coast of Australia. This was nothing unusual; people had been crossing back and forth between Australia and the fragmented bottom bit of Asia for millennia. What was unique about this particular landing was that there was a canine onboard the boat. This was the first time in fifty thousand years that a new species of placental mammal had arrived in Australia; the only others on the continent at that time were humans, bats, and rodents.[1] Nobody knows what that proto-dingo's role was on that voyage. She might have been a pet, or she might have been a ready supply of meat. Perhaps she was brought along as an item of trade, or she might even have been onboard to protect the crew from harmful spirits of the sea.[2] We don't even know whether bringing such creatures along on sea voyages was a regular occurrence, because the genetic diversity of contemporary dingoes is so limited that the entire species could well be descended from only a handful of individuals from as few as two such crossings.[3] All we know for certain is that this first arrival heralded the introduction of a canid species that would come to occupy the entire Australian mainland. This species upended the faunal composition of Australia and was complicit in the mainland extinction of two indigenous marsupial predators: the thylacine and the Tasmanian devil.[4]

Richard Francis argues that since their ancestors' arrival in Australia, dingoes have been undergoing a process of "reverse domestication."[5] His reasoning is based on the premise that they are descended from semidomesticated dogs of Southeast Asia and that since colonizing Australia they have become more reliant on the bush and less dependent on people. Of course, this argument is not without problems, as it implies a bidirectional path toward either a domesticated state or a wild state—a path so loaded with dualisms that it needs a double yellow line down the middle. It also fails to appreciate that evolution does not go backward. But it does have some coherence when it's fitted within a framework of ecological separation that I suggest defines domestication processes. If we consider domestication as separation from wider ecologies, then those first dingoes were well and truly on that path. In Southeast Asia, it seems that they had already been incorporated into human societies, and in this way they had been undergoing a process of separation from the ecological networks of their wolfish ancestors. Whether those semidomesticated dingoes were pets, livestock, commensals, or hunting companions, we don't know. But the evidence that the first Australian dingoes came to Australia with the assistance of humans tells us of a profound separation on a continental scale. What's more, it is likely that they were not just stowaways who went bush as soon as they hit dry land, because the archaeological record shows a long association with humans after their arrival. But we also know that dingoes as a discrete canid population underwent a process of ecological (re)connection with the Australian continent and its animals and plants, so in this way we could cut Francis some slack and suggest that, yes, they were undergoing a process of reverse domestication. Over the past four thousand years, dingoes have been redoing the undone.

In spite of their "reverse domestication," dingoes never really severed their connectedness to human camps, or, to be more precise, Aboriginal people never really let them. If the recent past is any sort of guide to the deep past, then we can be pretty certain that dingoes provided the original Australians with a ready supply of puppies. Why did these people want puppies? Well, there's some evidence that they were for eating. There's an account by a European explorer named Michael Durack who, when traveling in the Kimberley in 1900, happened upon an Aboriginal man preparing dinner. According to Durack, the man had two puppies roasting on the fire and another few he was keeping for a later meal.[6] And Norman Tindale has noted of the Pitjantjatjara in Central Australia that they practiced "increase ceremonies" in May that coincided with the whelping season, when they went in search of puppies to eat.[7] From the same region in 2007, anthropologist Diana Young was told by an informant of the best way to cook dingo puppies: cut open their stomachs, fill them with bloodwood leaves, and cook them in their skins in a pit oven.[8] But this and other accounts of dingoes as food are scarce in comparison to those that say the desire for dingo pups sprang from the need for an object of affection. During whelping season, people sought out dingo dens, whereupon they killed the bitches and took the puppies back to camp—not to eat, but to love. Debbie Rose was told of this process by older informants among the Yarralin people of Australia's Northern Territory. The position of the stars dictated when it was a suitable time to acquire pups. Dingo lairs were raided and newly weaned pups brought to camp to be cared for and socialized to the human group by specific owners. The pups grew up among humans, but as they matured into dingoes, the bush beckoned. For this reason, people sang certain songs to encourage the more amenable dingoes to stick around. As for the dingoes who stayed but failed to adjust, instead becoming a threat to people, these were fed with a poison made from an orchid.[9]

Annette Hamilton outlines the same puppy adoption process in her account of Jankuntjara people in Central Australia. The typical method was to track a dingo bitch to her den during the breeding season. The bitch was speared and the pups taken back to camp, apparently at the insistence of the children, although Hamilton suspects that adults had a stake in the game too. While other animal species brought to camp for the amusement of children were destined for the cooking fire as soon as the novelty wore off (or the limbs were torn off), dingo pups were effectively adopted. The pups were raised in camp and doted on by both children and adults. In fact, adult women were reported to carry dingoes around their waists, holding the front and rear paws in each hand, respectively, in the same manner as they carried children. This must have been delightful for the dingoes, who would otherwise collect burs and thorns in their footpads. The pups were given names, often with reference to mythology, and they were carefully groomed, hand fed, and rubbed with fat and ochre to protect them from spirits. Dingoes were even suckled in place of deceased children. On maturing, the dingoes were not subject to the same levels of affection, but no matter; they decamped.[10] The difficulty of getting food from humans, combined with dingoes' biological imperatives, returned them to the bush, and the cycle repeated itself. Hamilton calls it a self-regulating system, in which people could indulge in caring for pups without the consequent problems that accompany adult dogs.[11]

But as with so many of Australia's Aboriginal systems, this one was disrupted with the arrival of Europeans. After colonization, Jankuntjara people began acquiring European-derived dogs from settlers in the area. Like dingo pups, the European pups were also indulged, named, and subjected to loads of affection, but on reaching adulthood these dogs had no inclination to take their chances in the bush. Millennia of selection for domesticity pinned these Euro-dogs firmly to the company of humans, and so they stuck around, becoming, in Hamilton's words, "useless and dependent unfondlable adults."[12] The dogs chose for themselves a patch of camp territory and defended it against other dogs with whom they collectively and, according to Hamilton's experience, aggressively defended the entire camp from strangers. They bred prodigiously, to the point where they outnumbered people in camp and actively competed with people for food. Children soon learned to eat standing up with their hands held high, and storage of food required platforms to be built six feet above the ground. Despite constant discouragement by way of fists, sticks, and metals bars, camp dogs still consumed 10 percent of the available food supply.

The Jankuntjara people certainly recognized that the abundance of dogs was a problem, but at the suggestion that they cull the numbers, they expressed horror. "Do men kill their children when their own hands brought them up?"[13] Hamilton raises the question here: why, in the face of the acknowledged problems presented by an abundance of dogs, do people persist in acquiring puppies (of European origin)? The answer, according to the anthropologist, is the need to nurture. Hamilton presents two very telling charts, one showing a positive correlation between the age of Jankuntjara adults and the number of dogs kept, the other showing an inverse correlation between the number of resident children per person and the number of dogs kept. In other words, the older the adult, the more dogs he or she kept, and the fewer children people had, the more dogs they had. Of course, it could be the case that advancing age simply brings with it more opportunities to acquire dogs, but we should not discount the all too human need for close relationships and the rewarding feelings derived from pouring love onto infants, especially puppies.

In fact, Hamilton's case study might reveal much about domestication processes in general. A lot of theorizing has implicated hunting as the nexus around which humans and wolves converged and domesticated each other. This speculation always resonates with the defunct hypothesis that positioned hunting as the keystone of human evolution. The valorization of hunting, as both a male pursuit and a process mover, is a very androcentric view that historically appealed to male scholars because it places males squarely in the driver's seat in human evolutionary processes. But it certainly does not jibe with ethnographic observations of Aboriginal people, which show that dingoes had very limited utility in hunting because they were so unruly.[14] In fact, it was more productive to follow the tracks of a wild dingo who was chasing prey and take the carcass from her than it was to bring a tame one along on a hunting trip.[15] This is why, after colonization, Aboriginal people were very keen to acquire European dogs—because they could actually be helpful in hunting. So while I wouldn't write off entirely the utility of canids as hunting companions or guards in early canine domestication, let us take a page from Francis Galton and consider the agency of women and children and the all too human desire to nurture. Anathema as it is to the action-movie view of canid domestication, the process could well have been driven by hugs and cuddles rather than spear points and bleeding boars.[16]

What we know about the relationships between dingoes and Aboriginal Australians opens up a hitherto unexplored aspect of dog domestication: inhibited dispersal. In all of the theorizing about the original synthesis of humans and dogs, this element has been given no consideration at all. While pet keeping, hunting, guarding, and commensalism all offer up a reason for dogs and humans to become associated, none of them addresses an issue that is crucial to the process of domestication: getting the dogs to stick around. Most canids disperse when they reach maturity. This makes evolutionary sense in that it prevents inbreeding and overcrowding, but it seems to be something that has dwindled in dogs. As for the behavioral differences between dogs and wolves, most scientists concern themselves with animals' socialization to humans. An example of this is the detour task, in which dogs and wolves are faced with a transparent V-shaped barrier, behind which sits a food reward. Wolves and dingoes easily solve the task independently, while dogs look to the human experimenters for guidance. But the difference between wolf and dog dispersal tendencies has been utterly ignored. This neglect is due in no small part to the paradigm of scientific testing. Dogs and wolves subjected to behavioral experiments are prevented from dispersing whether they want to or not; researchers can't have their captives wandering off to other experiments. This also applies to the farm-fox experiment. The foxes were bred for socialization to humans, from which various doglike characteristics emerged, but the very nature of an experiment using captive foxes precluded any discoveries about inhibited dispersal. Yet this is something as crucial to canid domestication as any other trait, because if those proto-dogs had not stayed around camp during the early phase of domestication, the original Aboriginal Australian relationship with dingoes would be the global norm.

For Aboriginal people, dingoes were not just pets. Physiologically, dingoes bear a much closer resemblance to humans than to marsupials, so it's not surprising that they hold a prominent place in socio-cosmological terms. In Aboriginal accounts, Dingo is a creative force par excellence. He scratched out the landscape, creating rivers and water holes and, by extension, the uplands through which these features are carved. He gave birth to the first people and established the laws by which they should live, laying out the ground for relations between kin and strangers. He warned people of the dangers of the world. He even made an act of creation out of his own death. Debbie Rose recorded a story of Dingo and Moon that directly addresses the issues of life, death, and reproduction. In the dreamtime before there were humans, Moon offered to share with Dingo eternal life; Dingo would die with Moon—would go into the ground—and then reemerge on the fourth day. It sounded like a good deal, so Dingo followed Moon into death and darkness. After four days, Moon called to Dingo to follow him to rebirth, but dingoes are a recalcitrant lot. Dingo refused. For no apparent reason other than to disobey Moon, Dingo remained in the realm of the dead, and from that time bequeathed death to humankind.

The man who told Debbie this story was ruminating on his own mortality at the time, and he reflected ruefully on Dingo's decision, which consigned humans to mortality and death. But Debbie teased apart the story of Moon and Dingo to show that Dingo's actions in fact freed both him and humankind from what she calls a "form of obliteration."[17] For all his smugness, Moon, not Dingo, is the tragic figure in the story. Through eternal rebirth he is consigned to a fixity that makes his existence barren. He lives in a closed system where nothing is lost and nothing produced. There is no change but the extent of shadow across his face; after four days of death he returns: all remains the same. Dingo, by contrast, through choosing death, creates time itself and bequeaths to humans a world of possibilities and choices, of free will. His stubborn refusal to follow Moon gives us the freedom to create and to participate in life processes within a system where, ironically, the phases of the moon mark a world that is constantly coming into being. Death is a gift because it represents life, procreation, freedom, and immersion in the world. Thank you, Dingo.

Merryl Parker of the University of Tasmania collected more than fifty published Aboriginal dreaming stories, mostly from the Western Desert area, featuring Dingo as the protagonist. These origin stories, which describe a time called the Dreaming, make for fascinating reading and have much to say about how particular folks thought about dingoes. The dingoes featured are not generically modeled on a stereotypical dingo, although Parker picks up on one common character trait: recalcitrance. Even in stories, dingoes are difficult to tame. One story, "Peopling the Land," told in 1949–50 by Nipper Maragar and collected by Ronald Berndt, describes the metamorphosis of two dreaming beings into dingoes. Initially, they are called children, though this doesn't imply that they are human in form. Dreaming beings elude precise physical descriptions. As the story opens, these dreaming children, a boy and girl, are camping at Alawirwir near the mouth of the Alligator River. Their father brings a bagful of goannas and the children obediently start a fire for the evening meal. They eat up the goannas but the children are not satisfied. Despite their father's injunctions, the children proceed to gnaw on the goannas' bones. They begin to turn into dingoes. The father throws sticks at the unruly children, breaking the girl's arm and hitting the boy, who yelps like a dog. Mr. Maragar is a deft storyteller, and without being explicit, he punctuates the myth with subtle evidence of the children's theriomorphosis. As dingo children, the brother and sister run away from their father and travel through the landscape. These kinds of journeys are Aboriginal mythmaking meat and potatoes—they connect the contemporary landscape with dreaming beings, inscribing ancestral pathways into the land, fixing Aboriginal law into the immutable hills and crags. The two children dig out water holes along the way that stand as evidence not just of their journey but of their beneficence, for these water holes provide life for both people and animals. As dingoes, the children bear puppies all over the land, some of whom become rocks. But they also bear children and populate the land with humans, giving birth to the language groups Amurag, Winggu, and Mangeri. These ancestral dreaming beings are not categorically human, nor are they canid; they operate within the dreaming, where connections are made more explicit and rendered across individual beings. These dreaming figures not only inscribe the paths of ancestral law; they illustrate the relatedness of living beings and landscapes. As with Dingo, who gave us mortality and its blessings, we are indebted to our recalcitrant dingo children for their dreamtime escapade, because they gave life to the people and dingoes who care for the land. Their trail of water holes ends at Guwoid near Oenpelli.

There they dig a deep water hole and throw stones in every direction before diving down to the bottom for what is an exquisite ending. Declaring, "We are rainbows now," the two are transformed into the most powerful creative forces of the dreaming: rainbow serpents. And "those dogs" remain at the water hole to this day.[18]

It has taken a while, but colonial Australians have arrived at their own sort of veneration of the dingo, though this has come at the tail end of considerable animosity. Dingoes were at the forefront of the colonial encounter, as witnesses to the destruction of Aboriginal communities, sometime pets, and eventually targets of colonial eradication programs. In the account of the first gestural conversation between colonists and Aboriginal people, a dingo was present. Later, the first governor of New South Wales, Arthur Phillip, took a dingo as a pet. He was disappointed in his pet, describing him as "a very elegant animal, but fierce and cruel." Not the stuff of domestication. Nevertheless, the colonists saw the New World as raw material full of potential and entirely subject to possession, and dingoes were part of the package. In the 1830s it became fashionable to own a dingo, and many Sydneysiders became dingo fanciers. As pups, the dingoes were malleable, but as they matured they became unruly, killing neighbors' cats and chickens and confounding colonists' expectations of domesticity. The lives of pet dingoes were brief, and the fad had passed by the end of the century. This is not to say that dingoes were entirely useless to the colonizers. Dingoes rarely bark, and this characteristic is valuable in a cattle dog. Folklore has it that the Australian cattle dog, or blue heeler, was arrived at by adding a "spot of dingo blood" to Scottish collies.[19]

By the end of the nineteenth century, dingoes found themselves on the wrong side of the fence. The colonists had realized early on that inland Australia could carry a lot of sheep, so it wasn't long before vast herds of woolly grass eaters were making some pastoralists very wealthy. The problem was that of all the hoofed animals under human control, sheep are probably the easiest prey, and dingoes soon learned this. And so began a war that is still waged to this day. Colonists have tried as many means of dingo control as their imaginations will allow. They shot and killed the earliest sheep killers and seemed to be making inroads. In 1874, a letter to the editor of *Town and Country Journal* reported with some satisfaction that as a result of the killings, the dingo was "rapidly becoming extinct."[20] However, this report of the demise of dingoes was premature. Soon pastoralists were employing contractors to help out with the problem. So arose the occupation of "dogger"—an individual who makes a living shooting or trapping dingoes (and, today, wild dogs of any ancestry). Perceiving a threat to the mainstay of the newfound land's economy, governments got on board and began offering bounties on dingo scalps. The Queensland government's Marsupial Destruction Act, which facilitated bounties on kangaroos, was amended to include dingoes. A shot from a dogger's gun earned him five shillings at the time, a sum that had tripled by 1923. Eventually, dingoes' threat to sheep was seen as greater than that of marsupial competitors, and "dingo boards" were established to levy taxes and pay bounties. Killing dingoes became a lucrative business, and it wasn't long before the intention behind these measures was trumped by the rewards offered. Aboriginal people were quick to exploit the potential in the system: when they hunted dingoes for scalps to be traded for cash, they made sure to leave enough bitches to produce a continuing supply of pups for the next season. As the dingo numbers increased despite, or perhaps because of, the increasing bounties, white settlers complained, with some justification, that Aboriginal people were effectively "farming" dingoes.[21] Given the economic logic of the sustainable dingo harvest model, it was not long before colonial doggers got on board.

Needless to say, without an incentive to completely eradicate dingoes, the bounty system was unsustainable, and dingo boards began to go bankrupt. In the end, a royal commission was established to investigate the system and found it to be deeply flawed. Other methods of control were sought. The outcome was an ambitious structure that encapsulated the colonial project in Australia: a bloody big fence. Roland Breckwoldt devotes three chapters of his book on dingoes to this edifice, known as "the dingo fence," such is its importance in the story of dingoes and colonization. At 5,614 kilometers (3,488 miles), the fence is the longest in the world, and it was even longer prior to 1980, at 8,614 kilometers (5,352 miles). It follows a meandering pathway from South East Queensland up into the central west and then hugs the border of New South Wales before zigzagging through South Australia to the Great Australian Bight. It was originally a rabbit-proof fence between New South Wales and South Australia—two states that were convinced of the failure of the other's rabbit-control measures. But it fell into disrepair and only came back under consideration when sheep farming (and the threat of dingoes) spread west. At the time, people running cattle were unconcerned about dingoes, but a growing number of pastoralists determined to run sheep pressured the government to reconstruct and extend the rabbit-proof fence.

In the harsh climate and acidic soils of the outback, the fence required a lot of maintenance, but such was the determination to exclude dingoes that it had some success. Australia thus "rode the sheep's back" to economic success by keeping dingoes out.[22]

On through the twentieth century, the colonial imagination continually reconceptualized dingoes within the project of coming to terms with the upside-down land. Conservation was an unknown concept in the early days, so eradication by means of shooting, trapping, and baiting was at once desirable and seemingly within the means of modern man. By the 1930s, Tasmanian farmers had celebrated the passing of the last thylacine, and the thought that dingoes could follow was tantalizing indeed. For a sheep farmer, a land without dingoes was a step closer to an earthly heaven; for urbanites, dingoes didn't matter, although urban conceptions of them aligned neatly with the farmers' antipathies. To call someone a "dingo" was to call him a lying, cheating coward, and the epithet was thrown around in state and federal parliaments with howls for added effect. As Merryl Parker says, representations of dingoes during the early twentieth century had less to do with what dingoes were like than with what Australians wanted them to be.[23] Dingoes were cruel, skulking killers of economic success, and the men who trapped them, clubbed them to death, and stomped on their puppies were heroes of the Australian bush. But values are by nature changeable, and the post–Second World War era saw a shift in attitudes. Over time, the project of Europeanizing Australia gave way to an environmental discourse that instead sought to promote a unique Australian identity. Among the increasingly urbanized population, nativeness became de rigueur and Australia's econationalism came of age.[24] For dingoes, this meant an appreciation by colonists that they had not previously enjoyed. As with their verminous compatriots the kangaroos, dingoes were made emblematic of an authentic Australian "nature," symbolizing the Australian bush that marked the country as unique.

In 1980, when a dingo took baby Azaria Chamberlain from a tourist campsite, the public and the courts found it easier to blame the child's mother for her death than they did the dingo. As Parker says, the dingo of this era was seen as a larrikin, the canid compatriot of the stereotypical Australian male, and in this case an innocent animal accused of murder by an emotionally restrained woman of an unorthodox religion (Seventh Day Adventism).[25] Thereafter, on K'gari (Fraser Island) in Queensland, dingoes were portrayed as family-friendly icons of the Australian bush. Tourism brochures promoted close encounters with dingoes, and even though the National Parks Service disapproved, tourists encouraged these encounters by doling out food scraps around campsites. Urbanites visiting the island were sold on the idea of a cartoonish wild canid that was as amenable as a domestic dog; all they had to do to encounter one was step outside their tent with a sausage. As a result, the dingoes on K'gari became bolder and more aggressive. In 2001, among a series of incidents that should have raised red flags all over the place, nine-year-old Clinton Gage, who was enjoying a holiday on the island with his family, was killed and partially eaten by dingoes.[26] In response, the Queensland premier ordered a cull of the dingoes on the island, but public opposition soon arose, and the parents were blamed instead. But, as Parker observes, the parents had been duped. They were sold a line on endearing, iconic dingoes in a family-friendly vacation spot, a line dangerously at odds with reality.[27]

The dingo has since wandered into a complex set of conceptions in Australian imaginations, and the waters have been muddied the more since they have interbred extensively with domestic dogs. For sheep farmers, they are still the pests they've always been, and the hybridization issue gives an added justification for exterminating them: if the sheep-killing canids are not "pure" dingoes, then they don't have the right to life that native animals do. Some conservationists view the hybridity issue as the problem, as they see dingoes threatened with extinction through interbreeding with feral dogs. Dingo ecologist Laurie Corbett predicted twenty-five years ago that dingoes in the wild will be extinct by 2100, not through persecution but by hybridization.[28] It is not car parks, poisons, or guns that will eradicate dingoes but hybrid vigor. Even though there's nothing intrinsically maladaptive in hybridity, for various reasons the genetic purity of dingoes matters to many Australians, and they reason that a dingo and a wild dog are not equals, even if they perform the same ecological role.[29] Fiona Probyn-Rapsey finds the conflation of hybridization and extinction very odd indeed. Shouldn't extinction be associated with death and disappearance rather than with sex and reproduction? She finds the hybridity issue an extension of the colonial fixation on racial purity and hybridity in humans, which characterized racist policies of the early twentieth century.[30] According to Fiona, dingoes are marked by the same public attitudes that saw children of mixed Aboriginal and European parentage taken from their (Aboriginal) mothers and sent to schools to be "whitened." By incorporating them into colonial Australian society, over generations, the Aboriginality would be bred out of them. Mother-child bonds were severed—they were secondary to the need for a racially ordered Australia—and an entire generation was stolen. Hybridization fosters metaphysical panic in people who see culture through dualistic frames of reference. Hybridity makes it impossible to say whether a person is one thing or another; it undermines superficiality and drags people kicking and screaming into a world where race becomes meaningless and the wonderful truth—that the world is complicated—slaps them in the face. The same panic over hybridity infects the view of dingoes, and very pedantic arguments are put forward to make this panic seem justified. One of these is that domestic dogs can breed in any season and produce two litters per year, compared with the comparatively brief breeding season and single litters of dingoes. According to this reasoning, hybrids will breed unhindered and the country will be overrun by wild dogs. Its exponents don't consider whether this is adaptive, as though genes rather than environments constrain breeding. So within this discourse, dingoes are threatened and rare, which immediately translates into value. This extends to the dingo pet trade, where tests of genetic purity determine the value of the pet. The higher the percentage of "dingo genes" in the test result, the purer—and more desirable—the dingo.

For some people, the concern with conserving genetically pure dingoes is a powerful motivator. I'm visiting the Australian Dingo Sanctuary in Victoria, where the owner, Gwen Thornton, and a squad of volunteers maintain a colony of captive purebred dingoes. Gwen hails from a dog-breeding background as an internationally recognized judge and breeder of 150 champion pedigreed dogs. But her passion now lies with dingoes. In the early days, she wasn't able to keep dingoes because of government legislation, but she worked to influence legislative change in Victoria so that she and others could. When the government got on board, she was involved in setting the guidelines for keeping dingoes in captivity. She prescribed the minimum thirty-square-meter floor area and 2.4-meter-high fence (since increased to three meters) of a dingo enclosure. Now she keeps a colony of "pure" dingoes, which she breeds for zoos, sanctuaries, and private owners. Each dingo undergoes a genetic test to determine purity.[31] Gwen is working with geneticists to increase the number of markers by which dingo purity can be measured, which will enable her to refine the breeding program. She hopes that one day the conditions will be right for her pure dingoes to be released into the wild. As things stand, though, the current abundance of wild hybrids would make such a project pointless.

Gwen greets me at the door of her country home and takes me around back to the dingo enclosure. On the way, we pass through a large shed that serves as a visitor center, where posters of dingoes adorn the walls. Down a path from the visitor center is a very large shed that is partitioned into eighteen regulation-size pens, most of which hold two dingoes. At either end of this structure there are two fenced paddocks that serve as exercise yards. The pens face east-west to avoid the harsh winter chill and hot summer winds, and the rear half of each pen is sheltered by the roof of the shed. The unroofed sections have wire mesh ceilings to prevent escapes, and the service corridors in the front have gates at each end. Dingoes are consummate escape artists, so these measures are not superfluous. As we walk past the pens, the dingoes show an interest in me, but no one runs to the front of the pens to greet me. No one, that is, except the newest litter of pups. Five little dingoes—three ginger and two white—crowd around the gate, brimming with excitement. Gwen and one of her volunteer staff herd them out of the pen and we head for the exercise yard, where Mum and Dad dingo are waiting. The pups explode into the open area and proceed to pounce on everything—parents, fallen logs, Gwen and myself. In total contrast to their parents, who keep their distance and eye me suspiciously, the pups are all over me, standing on my lap, jumping on my shoulders, and licking my face and ears. It's like seeing two different species—the parents as wild dingoes and the pups as domestic dogs. But, as Gwen tells me, this will all change when they are about sixteen weeks old. The pups will lose their hypersocial boisterousness and take on the aloof dispositions of the adults. This won't preclude relationships with humans—they will bond with someone early on and stay bonded for life—but it will keep them from licking the ears of strange humans. Still, there is one thing unusual about the pups: they were born in November, about three months after whelping season. If these unseasonal pups were born in the wild, the Australian bush might not be as accommodating as the dingo sanctuary, and dingo behavioral ecology suggests that their chances of making it in the wild would be slim. As it is, they will be cared for. Food, shelter, and protection from parasites will all but guarantee their survival, and Gwen informs me that they already have zoos, sanctuaries, and private homes ready to take them on. The new owners will have to sign an agreement that the pups will never be bred with anything other than pure dingoes. So there's a chance that these pups will pass their unseasonality on to another generation and into the gene pool of captive-bred dingoes.

I ask Gwen about the criteria for choosing pairs of dingoes to breed. "When you're breeding show dogs, you're breeding phenotype, you're breeding morphology; you're not breeding instinct and temperament and nature," she replies. "So when you move to something like a dingo, which is so much more comprehensive, that is nature's animal, nothing to do with me, you realize that you have absolutely no right to meddle with these animals' genetics other than to make sure they are healthy breeding animals and that they like each other. I look for stability of nature, but I have no control over that. There's a natural range of temperaments, and nature wants that range. Unless you have a range, you can't adapt, and so keeping that diversity in their breeding lines is important. I'm always looking for new bloodlines." Is compatibility an issue in breeding them? "If they don't form a bond, then there's not going to be a conception, but if there is, then the partner is likely to kill the young. The female dingo is the alpha, make no mistake, so unless the male complies there won't be any bonding or breeding. There won't be any conception. They're so different to domestic dogs. Nowadays, with show dogs, you get the semen from overseas and inseminate the bitch. She has the puppies and you look after them. What happens here has got to be as close to nature as humanly possible, given that all of the animals are predestined to go to various places where they'll have to interact with humans." If they were destined for wild release, they'd be brought up differently, but as it is the pups are habituated to human contact from the get-go and primed for a close relationship with a human caregiver. I ask Gwen if many dingoes come back from people who've failed as dingo owners. "No," she says. "It's always been my view that they have to know beforehand whether they'll fail. I make sure that people know that this is a marriage. It's a twenty-year marriage. If you want a dingo out of here, then you need to understand that it's going to bond only with you. You can't put it in a boarding kennel and go away. It's not something you can try out and pass on to someone else. It's a rare dingo that will bond with somebody else." Do you get a lot of people coming here wanting dingoes as pets? "I think I do. A lot of people who come here are too frightened to ask once they've listened to me. But most of them go home with a very different idea. They get the truth: that dingoes do not make good pets."

I'm not normally one for breakfast TV, but such is the information age that I'm quick to find out about an interview on Australia's *Today* show with someone promoting dingoes as pets. Watching the replay, I see two predictably attractive white presenters, each holding and petting a dingo pup. The male presenter says, "If you're thinking of adopting a pet to your household, you might consider dogs, cats, fish. But what about a dingo?" The other presenter interjects, "They're very cute." She introduces us to Charlie from Sydney Dingo Rescue, who is "on a mission" to find homes for rescued dingoes. Charlie is cuddling a slightly older pup. Our presenter gets right to the point. "Look, there's a bad rep for dingoes for obvious reasons—that they're a wild animal, that they're violent (as though dogs are not). Is that the case?" Charlie offers a tactful response, saying that dingoes are certainly "unique" but that in the right environment and with the right caregivers, they can make very affectionate, loyal, intelligent animals and companions. "They can make good pets?" While Charlie says that he would like to see more dingoes in the wild, they do make good companion animals. They are trainable, but they're very smart, so they need a lot of enrichment and a lot of stimulation. At this, the presenter suggests, "Almost like a border collie." Charlie concurs. "Very high-energy, very smart creatures." Given their recent history of attacks on children, the next question alludes to safety. "I've got a young family. Is it good to have a dingo as a pet for a family?" Charlie is reassuring. He says that he knows lots of dingo owners with kids, and that with the right environment and the right relationship, they are fine. But you need to have a backyard and high fencing. As far as food goes, they don't digest grains or starch too well, so raw meat is the way to go. But Charlie says a raw diet is good for any canine. And colors? Charlie dispels the misconception that dingoes only come in the ginger/white coloration. They come in colors ranging from white to black-and-tan. Any viewers out there worried that a dingo will clash with their gazebo can rest assured.

Unsurprisingly, after Charlie's appearance with the puppies on the show, he is flooded with calls from people wanting to adopt a dingo. I email Charlie and ask if he wants to talk to me about the domestication of dingoes. He responds, "We don't support the keeping of native animals or any wild animals for that matter, as pets, and in an ideal world I would like to see dingoes only in the wild and not as pets. We keep them in captivity now only because there is no other option." There's also a lot of discussion on social media about the way Charlie promoted dingoes on the show. Most of the comments on *Today*'s Facebook page are from people sharing their own experiences of love and happiness with pet dingoes. Still, a lot of them mention that their dingoes are crossed with various breeds. The photos they post depict a hodgepodge of pets with floppy ears, dappled coats, and bulbous Chihuahua eyes. But the comments overflow with love and happy memories, even when the dingo has proved to be a difficult pet. One woman who had a "beautiful dingo cross" whom she "loved dearly" also calls her the "worst dog ever." Apparently, the dog was an escape artist and needed a concrete enclosure. But among the happy anecdotes, some people offer sober advice: "Not everyone can own a dingo, they simply don't know how to accommodate one." "Dingoes make great pets but you really need to consider if your home is suitable for one." Most of the comments in this vein are concerned with the needs of dingoes in terms of space, enrichment, and constant attention. And then there are those who are simply dead-set against keeping dingoes as pets, no matter the circumstances: "Shame on the Today Show. You should do your research before bringing inexperienced people on national TV stating that 'dingoes make great pets.' But no, you wanted a cute and cuddly story that will bring more pain and destruction to our unique and iconic *Canis dingo*." Another viewer is less diplomatic: "Stupidest idea I've ever heard of. They are wild dogs don't try to domesticate them. Stop trying to save the world and let nature be." Charlie responds to the criticisms and the admonishment of a dog trainer and animal behavior expert on the ABC News website. He says, "Ultimately, we'd like to see more dingoes in the wild and less in captivity, but for the ones that are abandoned, it's about finding them a suitable owner. If a dingo home is not working out, we are always prepared to take back the dingo for the welfare of the dingo."

So are dingoes really difficult pets? At a 1999 "Symposium on the Dingo," Barry Oakman of the Australian Dingo Conservation Association was unequivocal: "[Dingoes] make lousy pets when compared to domestic dogs." Why? Because dingoes lack the capacity to reason. "This animal is, by nature, governed by instincts in every situation, be it survival, mating, whelping, hunting, play, and dominance. Even in captivity it will retain these true wild traits." For Oakman, dingoes' insecurity and low tolerance for fear lie at the heart of the problem. This is because fear and insecurity can easily lead to aggression or flight. Dingoes are also such effective escape artists that keeping them in an enclosure is extremely difficult. If one does get out, his strong prey drive will soon direct him at the neighbor's cat. Still, what Oakman refers to as being governed by instincts might be a subjective perspective based on a comparison with domestic dogs whose instincts compel them to defer to humans even when they feel unsafe. Perhaps the fact that dingoes make decisions independent of humans gives Oakman this impression, and perhaps this is the core of the dingo pet problem.

My dear friend Liz Marshall Thomas had a dingo named Viva to whom she was very close. Viva was very much a puppy out of place. Liz acquired her from a puppy mill near Chicago where someone raised dingoes. She wasn't impressed with the breeder and would have liked to take all the puppies away but ended up with Viva alone. Liz loved Viva and her love was returned. Viva was different from Liz's other dogs; she used her paws a lot to manipulate objects and was more aloof, both quintessential dingo traits. Unlike the other dogs, she showed no interest in human visitors, and she certainly didn't jump into people's laps. She didn't seem to be afraid of people; she just wasn't that into them. According to Liz, "Viva would stay back, keep watch on all that went on, but not interact with them. She might let them touch her, but she'd pull back a little when they tried." As for being intractable, Liz never really gave Viva the opportunity. As she said of all her dogs, "I'd rather learn from them than for them to learn from me. I like to see what they want to do, what they find interesting." Liz makes no demands on her dogs, so there is no opportunity for them to be disobedient. She might ask things of them, such as to come along with her, and in this Viva was not as quick to respond as other dogs, but in no way did Liz consider this recalcitrance. "All in all, she was more aloof, but just in her demeanor. She didn't feel aloof, it was just the way she acted. She didn't feel she needed to prove anything. I'd say she accepted our relationship as it was."

For wildlife consultant Roland Breckwoldt, dingoes are amenable pets. When he and photographer Gary Steer traveled Australia in search of dingoes to film, they brought along one of their own from Breckwoldt's farm, a dingo called Maliki. Breckwoldt says it was necessary to bring Maliki along because of the constant care and attention needed to keep him "quiet and responsive," as though domestication started anew every dingo day. During their travels, Breckwoldt found Maliki more help than hindrance, as people gravitated toward the native canine and poured forth their own stories of dingo encounters. There was one common thread among bushies and pastoralists who took on young dingoes and tried to make them into useful working dogs. Invariably, the dingoes were shot as adults because their utility was outweighed by their tendency to hunt anything that moved. Yet Breckwoldt made no such complaint. He only ever chained Maliki when they were trying to film wild dingoes, because of the tension that arose between Maliki and his wild cousins—something that, curiously, was never apparent in Maliki's encounters with domestic dogs. It would seem that humans are not the only species to make a distinction between wild and domestic.

The famous Austrian psychologist and ethologist Konrad Lorenz remarks on this distinction as a different expression of two canine traits: the bond that ties a young pup to his mother and the loyalty of a wolf to his pack. Lorenz is misled by his own suppositions here, in that without the benefit of molecular biology he divides the dogs into two lineages: wolf dogs and jackal dogs. He reasons that the jackal-derived dogs are not as loyal because their ancestors did not develop the bonds of the pack. But misclassification aside, his general premise—that the infantile bond of wolf pup to mother is retained and redirected onto his human master—is not without credence. The submissiveness, inhibited dispersal, and physiologically neotenous traits of domestic dogs are the very things that distinguish them from dingoes—along with a few physical limitations and reproductive overexpressions. In effect, this tells us that the famed recalcitrance of dingoes is the marker of a markedly adult canid. In this, Lorenz is also an expert, having had a pet dingo in his home in Austria. As a pup, the dingo showed all of the submissive obedience of a domestic dog, but on maturing he entirely seized his independence. This is not to say that he lost affection for Lorenz—the two remained close—but, as Lorenz says, submission and obedience played no part in their feelings toward each other. So the question of whether dingoes make amenable pets seems to have little to do with the dingoes and a lot to do with human expectations and behavior. If someone loads her dingo with a bunch of expectations cut from a pup-like lapdog, then there is little doubt that the dingo will be a difficult pet. It's unlikely he'll display the blind loyalty, trust, obedience, and unconditional devotion of a sycophantic domestic canine. But the success stories show us that some kind of mutual relationship is possible; it just requires that the human partner be interested in and willing to learn about the adult dingo's way of being in the world.

As I drive from rolling farmland into forest, the mist vanishes, as though the trees have sucked every drop of moisture from the air. Technically, it was cloud that shrouded the pasture, making apparitions of the cows and scattered trees. I've driven up the escarpment and onto the New England Tablelands, where cloud base has descended on the landscape. But the forest I've just entered is utterly clear, and I can see farther through the trees than I could across the open pasture. It's only a ninety-kilometer drive to where I'm going, but it has taken a couple of hours. The road undulates, zigzags, and weaves, and the last ten kilometers of the journey are unpaved. I also stopped on the way to pick up some milk and bones. The milk is for my friend Margi, the bones for Yindi, Mojo, and Dawa—her dingoes. At a bend in what remains of the road, I slow down. If I didn't know the track, I'd miss the turn, as I did the first time I came this way. It's the tiniest of gaps between the trees, and to an unwitting bystander it would seem that I'd gone mad and veered off the road into the forest. The rough, overgrown track is the sort that's normally accessible only to intrepid goats. It leads me a couple of kilometers down to a rocky creek crossing, where I stop at a strip of electrified tape. I get out of the car and my senses are assaulted. The chill and clarity of the air and the sounds of the forest give me pause. A chorus of kookaburras rings out across the forest, their collective laughter probably a response to the arrival of my car. I have the utmost respect for these birds, and not just because they have the temerity to kill and eat Australia's most poisonous snakes. They're highly family oriented. At the rising and setting of the sun each day, kookaburra families gather in significant trees bordering their territories and ring out their loud, laughing calls. They also do this in times of stress, and even when a young kookaburra fledges. Amid the cacophony of laughing kookaburras, I can hear another sound through the trees—the bark-coughs of the dingoes alerting Margi to my arrival. This forest holds few secrets. I lift the insulated end of the electrified tape to unhook it, cast it aside, get back in the car, move past it, and then get out and reattach the tape. I drive across the creek and along the last bit of track before emerging from the forest in a small clearing in front of Margi's cabin.

Mojo is the first to greet me as I get out of the car. He's as much dog as he is dingo—his father was a wild dog and his mother was a kelpie—and it shows. Mojo doesn't have the flexibility of a dingo; he can't turn his head 180 degrees and he can't rotate his paws like a dingo. His attempts at howling are also pretty comical. As if to underline his doggyness, Mojo wags his tail and makes a show of greeting me. Yindi, who is much more dingo than dog, stands in front of the veranda, simply observing my arrival. I hold out my hand for her to sniff. Meanwhile, Dawa, who is very much a dingo, watches me from the enclosure behind Margi's house. Margi steps off the veranda and gives me a hug. She has graying hair and a few tattoos. Standing in front of the isolated cabin in her cargo pants and singlet, she looks like an unarmed preppie waiting for society to unravel. But this self-identified pagan lives in this remote place not out of paranoia but out of a stronger sense of connection to the bush and her dingoes than to society. Margi calls off Mojo with some words in Gumbaynggirr, the local Aboriginal language she's been learning.

"Come inside and see her," she says, and I take off my boots on the veranda and step inside. The cabin is dark, dusty, earthy smelling. The inside feels little different from the outside, and I imagine how cold it would be in winter. The little woodstove must get a workout at that time of year. There is an unfolded sofa bed in the middle of the living room among the piles of books and paper. This is where Margi and the dingoes hang out in the evenings. Between the sofa bed and the door is a cage about half a meter square. Inside is a water dish, a litter tray, and a dingo pup. This is Pippi Longstocking, whom I've come especially to see. Pippi is about four months old. She's curled up in the cage, so it's hard to see how big she is, but she's obviously a pup. She has a white forefoot, which is how she earned her name. She's afraid of me, so I make a point of sitting at a distance and not looking at her while Margi makes me a cup of ginger tea. I always feel carsick after driving the winding roads to get here, so the tea is more than just refreshment.

For Margi, dingoes are a passion and a life's work. She's had many on her remote patch of land and has loved them all (except perhaps for one overly aggressive dingo who had been allowed to dominate the household from which she got him). And through her organization, Dingo Conservation Solutions, she's extended her love to dingoes she'll never know personally. Margi has been training dogs since she was eleven years old, and while she still loves dogs, she is utterly infatuated with dingoes. Most of her dingoes were acquired as rescues, either orphaned, wild-born pups, or pets who haven't worked out for their owners. So it was serendipitous, given Margi's vocation, that Pippi came to her from the wild. On the day Pippi showed up, Margi was away; a friend, Gary, was looking after her dingoes. He was awakened in the early morning by a commotion outside the house. He ran outside to find Yindi and Dawa, the fuller-blooded dingoes, attacking an unfamiliar pup: Pippi. Gary intervened to save Pippi, but Yindi, the alpha bitch, didn't want him to have her and was pretty aggressive about it. Nevertheless, he managed to throw a blanket over the pup and bring her inside. Margi arrived home that afternoon and found Pippi hiding in the spare room. She used a noose pole to keep Pippi still while she grabbed her by the scruff of the neck and put her in a cage. "She was emaciated," Margi tells me. "Literally skin and bone. Her hair was short and she had lots and lots of ticks on her; totally covered in ticks. Big ones, little ones, shellbacks. She wasn't showing any sign of actual tick fever, but she was covered." Margi used pyrethrum to get rid of the ticks, wormed her, and gave her some soft food mixed with water. She didn't take her to the vet because she thought the stress might make her worse. She just kept her in the cage in the corner of the room and gave her time to recover.

Margi began the process of habituating Pippi by keeping her completely hidden with a zip-up cover. That way, Pippi could get used to human voices but feel less exposed. In Margi's experience, this is better than keeping dingoes in an outside enclosure. "When they're outside, their whole focus is on escape; it's all they want to do is get out there into the bush. They just refuse to accept their current situation, they'll pull until their teeth come out, they'll rip the wire. If you've got electric dog fencing, they'll get done by that, but once they've broken it to the point where it's not working they'll rip it out. They'll dig, they'll climb, they'll do anything." Day by day, Margi pulled the cover back a few centimeters, so that after three weeks, half of the cage was exposed. I go over to the cage and see Pippi curled up at the back. She eyes me warily but shows no sign of panic. She looks drained. "I found that when they're in the house, the cage is the safest place to be, so they don't want to escape because it's scarier outside of it. Because there's no bush. So now she doesn't want to go out of the cage, which is unusual at this stage. Normally they're out and about and checking out the whole house and destroying things and getting in all sorts of trouble. Pippi seems to be particularly not confident." Margi thinks this is a combination of Pippi's young age and a high degree of dingo blood. "I've had ones that were older—that were more confident—but they were hybrid. I'm yet to get DNA done on her, but I wouldn't be surprised if she's pure. There's absolutely nothing that I've seen so far that points to any hybridization. It may be that because she's pure she doesn't have those doggy genes that say 'people are okay' that the other ones might have had. The DNA test will tell."

I ask Margi why she thinks Pippi showed up at the house when she did. Surely she knew it was dangerous because of the other dingoes here. Margi tells me about the "black pack." This is a wild dingo pack of mostly black-and-tan individuals whom she sees every now and again, though fleetingly. She even calls to them by going to the hill behind the house and howling along with Yindi and Dawa. Mojo joins in with his own groans and woofs. Unlike Mojo, Margi is a pretty adept howler, with a variety of howls that she uses in different contexts. The black pack usually responds to Margi and her pack with a chorus of their own howls. This is how dingo packs manage to avoid one another: by letting other packs know where they are. But it's not all cordial between Margi's pack and the blacks. "One day, I heard my guys go off. They were out of the compound. I ran outside to see my guys running down the driveway toward the creek crossing here. The black pack were coming down to the creek crossing on the other side with murder in their eyes. I just raced down there with my arms above my head shouting like a Yowie. I just went 'Rah! Rah!' and they took one look at me and went *kachoong*, and they were gone."[32] Margi explains that on the day Pippi showed up there was a wallaby carcass at the bottom of the driveway. This must have attracted Pippi and her mother. "After Gary rescued Pippi, he called me, saying, 'What do I do, I've got this little pup?' While I was talking, I heard a howl and I said, 'Who's that?' Gary said, 'Well, it's not one of ours because they're all locked up.' I said, 'It's coming from over the creek.' So obviously Pippi had come with an adult, probably the mother. At the creek the mother must've thought, *I'm not going up there, that bloody scary woman's up there*, so she stopped. While Pippi thought, *I'm starving, I can smell carcass*, so she kept going and got nailed by Yindi and Dawa. So that says to me that there is a mother out there, because it sounded like an adult howl, not a puppy howl."

There's also a sibling out there. Pippi is small for her age, and her lack of assertiveness leads Margi to think that she's the runt of her litter. So what of her siblings? Why weren't more pups from the black pack caught near the carcass at the bottom of the driveway? Three weeks after Pippi showed up, Margi spotted a similar-looking pup nosing around the side of the house. He was a bit lighter in color and larger, but like his sister he had white socks. Margi reasoned that it must be Pippi's brother because the chances of a pup from another pack showing up in the black pack territory are practically nil, and the chances of another black pack female breeding and her pups being allowed to survive are also slim. The alpha bitch of a dingo pack is usually the only one who breeds, and she will kill the pups of other female pack members, even if they are her grandchildren. It's also interesting that Yindi and Mojo spotted Pippi's brother but didn't attack him (Dawa was in the compound). Margi thinks that her dingoes assumed that the pup was Pippi, whom they now know they're not allowed to attack. Considering Pippi's need for some dingo companionship and, when she eventually comes out of her shell, a playmate, Margi has toyed with the idea of catching her brother and bringing him home. "I think, wouldn't it be wonderful if I could get her brother too and stick him in there? It would cheer her up. But then I wouldn't want to do that to her brother either. He's surviving at the moment. Hopefully he's got parents out there who are feeding him."

Margi's reluctance to catch Pippi's brother derives from what she values most in dingoes: their wildness. While genetic purity is a way of rationalizing degrees of dingo traits or advancing the cause of dingo conservation, it is their wildness that draws her to these animals. Like a dingo, she sees herself as a free spirit. This, she believes, is why she has a history of failed relationships with men: the very wildness that attracts them to her is what they seek yet fail to tame. Margi is nobody's dishwasher. And so she nurtures the wildness in her dingoes. Rather than keep them in pens with high fences, she treks with her dingoes in the forest. Among the trees they encounter snakes, detect the scents of other dingoes, and hunt goannas and wallabies. "I feel for the wallabies, knowing they must have a family, but at the same time a part of me wants the dingoes to catch them," she says. The dingoes oblige and often bring home food enough to save Margi buying a week's kibble. So the idea of penning a dingo in a regulation-size enclosure with three-meter walls is anathema to Margi. Hence her response when someone who offered to house a dingo with her expressed concerns about the inadequate fencing. She told the person that if she was a dingo she'd rather be dead than cut off from the bush. If such freedoms mean that her dingoes are more likely to be bitten by snakes, get into fights, get shot, or be poisoned, then so be it. Genetic purity is less important to Margi than a dingo's connection to the bush, and the latter must be nurtured. This is why, in spite of having fallen in love with Pippi, Margi is mentally prepared to release the pup back into the bush to reunite with the black pack. If Pippi doesn't "tame down," then she would rather set her free than have her cowering at the back of a cage inside the house. Rather than catching Pippi's brother to make life better for Pippi, she thinks life for the little pup might be better spent with her brother in the bush.

Four weeks after my first encounter with Pippi, I once again make the trip to Margi's house. It's a lot warmer than last time—such is the changeable weather in the mountains—but I still need the ginger tea. I'm sitting on a stool in front of Margi's kitchen bench. Behind me, in the corner of the living room, stands an empty cage. Pippi is gone. She had been making a lot of progress, Margi tells me. The soft diet worked and she'd put on quite a bit of weight. As Margi planned, she pulled back the cage cover a little every day, so that Pippi could become habituated to the house. Then she opened the cage door, giving Pippi a chance to explore, though only in the evenings when the other dingoes were in the house. Pippi's nose-to-nose interactions with Yindi, the alpha bitch, were tense, but Margi intervened to make sure they didn't escalate into violence. Dawa was accepting and played with Pippi in the house until he got bored, and Mojo was quite happy to have Pippi in the pack. Surprisingly, at least to me, the animal who gave Pippi the most attention was Talon, the cat. He was the first to accept Pippi, and the two played together into the night. Talon plays rough, so he's not intimidated by the much larger dingoes, constantly launching attacks on whomever he can catch off guard. A turning point in Pippi's habituation came when Margi took her to the vet in town. On the vet's exam table, Pippi was terrified—she defecated all over the place. But when she got home again, she was transformed. She nuzzled Margi and played with the other dingoes; it really seemed like she was glad to be home. Margi thought things were going well. Perhaps that was why she wasn't careful about properly latching the front door. Margi was preoccupied, getting ready to go out; the cage door was open. Pippi seized the opportunity, escaping out the front door and into the bush. Margi saw her the next day at the back of the house. Mojo had brought home a dead wallaby, which probably explained Pippi's presence. Yindi and Mojo alerted Margi that she was back, whereupon Margi ran to the back of the house, only to see a flash of white foot disappearing into the trees. She intended to recapture Pippi, but a few days later she had human visitors, who camped in the clearing between the house and the creek. The people made a lot of noise, and Margi never saw Pippi again. Such is the call of the wild.

While Pippi remained in the bush, her DNA tests came in. Pippi's DNA was amplified at twenty-three satellite markers that distinguish dingoes from dogs; the results were averaged and showed 76 percent, with a conservative error of plus or minus ten percentage points. In other words, Pippi was three-quarters dingo, with some domestic dog ancestry. This surprised Margi, because she'd seen no indication of dog ancestry in Pippi. But it didn't change anything. Pippi's desire for freedom in the bush far outweighed any domestic tendencies, including any attachment she might have had to Margi, Talon, or the other house dingoes. Margi respected that choice and simply expressed her hope that Pippi would thrive. But there's an acerbic irony in Pippi's story. In classical trap-tripping, bullet-dodging dingo style, Pippi unwittingly exposed an Australian folly. A "pure" captive dingo is an oxymoron. Certainly, the dingoes now being bred for an unforeseeable future release into a hybrid-free Australian bush will retain their ancestors' physiologies. They will be expert climbers, physically flexible, and have a strong prey drive. But is that all there is to being a dingo? Are they not lethal hunters of wallabies? Do they not glide through the bush and disappear without a sign? Do they not occupy vast territories and warn other dingoes of their presence? Genetically "pure" dingoes are being domesticated. They are being swept up in the wave of unmaking, removed and separated from the ecological networks that created and sustained them. Wedge-tailed eagles won't scoop up their puppies; shellback ticks won't suck their blood; they will never kill a goanna or dodge a snake; worming tablets and flea collars will keep them healthy; a ready supply of food will keep them idle. For many dingoes, this might be an enviable life, but it is still domestication. It is a separation from the complex ecologies that enmesh dingoes and other species in the Australian bush.[33] The future ecologies of "pure" dingoes will be bounded by fenced enclosures and contracts ensuring genetic purity. The bush will make of them no demands, and they will make none of the bush. And if changes along the lines of the farm-fox experiment occur in this population, they will not simply be due to socialization to humans. The harsh master, the Australian bush, will have no say in whether pups born out of season will survive. Living among humans, alpha bitches will not be permitted to kill valuable pups of subordinate females. Indeed, breeding more than once a year will be a boon rather than a burden, such is the demand for pure dingo puppies. Sociability in terms of immaturity and sycophancy will be the preferred currency among dingo domesticates, and, genetically pure or not, something very akin to a dog is likely to emerge from this experiment. Consider the lovable pups born in November at the dingo sanctuary. Their futures lie in the human world; they will grow up in sanctuaries, zoos, or households where they will mature, and perhaps some will produce the next generation of purebloods. Meanwhile, in the bush, where genetically pure dingoes are disappearing, novel sorts of connectivities are being established. The irony is that domestic dogs are the progenitors of the wild things who will replace the now captive population of "pure" dingoes. It's a classic switcheroo: dingoes are going domestic, while the hybrid progeny of domestic dogs are going bush. They are reconnecting where once there was separation. They are inserting themselves into the ecologies of their dingo cousins and subjecting themselves and the Australian bush to an ecological paradigm that accommodates a twenty-five-kilogram canine predator. They are a product not just of wild sex among dingoes and dogs, but of a harsher world in which humans threaten their pups rather than cuddle them. While it might be inappropriate to call these new creatures dingoes—not just because of their hybridity but because they are not being domesticated like the pure ones—they are still occupying the former niche of the dingoes. Pippi Longstocking's DNA test came back at 76 percent dingo. She increased the percentage by other measures when she escaped through Margi's front door.

3.

Stingless Bees

What do you suppose?
A bee sat on my nose
Then what do you think?
He gave me a wink
And said, "I beg your pardon,
I thought you were the garden."
—ENGLISH RHYME

Of all relationships between humans and other animals, that which we have with bees could well be the most ancient—predating even the emergence of our bipedal ancestors. This is so because of the apian capacity to produce honey and the hominine propensity to devour it. Anthropologist Alyssa Crittenden goes so far as to say that honey consumption was crucial to human evolution as a source of energy and as food for ever hungrier brains.[1] Crittenden draws on evidence of other primates' honey-getting capabilities to argue that as early as 2.4 million years ago, *Homo habilis* were using rudimentary stone tools to open up hives and consume their contents. This was a key period of human evolution during which our ancestors adapted themselves to more open environments with novel challenges (including living in larger, more cohesive groups), and these demanded greater cognitive capabilities. So to argue that honey and stone tools constitute an adaptive niche is to argue that honey is in many respects responsible for the way humans are today. It is to bind bees and humans into a complex evolutionary story.

Primatologist Richard Wrangham agrees that our ancestors ate a bit of honey but disagrees about its importance to human evolution. Wrangham argues that our ancestors needed to be able to control fire before they could utilize honey to any great degree, and so honey is simply one of many resources that ancestral humans made use of. He doesn't doubt that honey was food for our ancestors, and indeed he shows that many other primates enjoy it, but Wrangham argues that "the inability of australopithecines to quieten honeybees would have been the critical constraint on their collecting large amounts of honey."[2] By "quietening," Wrangham means using smoke. The injection of smoke into the hive convinces the bees that their hive is about to be burned to the ground, so they prepare to evacuate; they gorge themselves on their honey stores and become inebriated, leaving the hive undefended and easy pickings for a *Homo erectus* wielding a sharpened stick, or a man in a veiled hat wielding a hive tool.

But Wrangham, with his emphasis on heavily defended hives and the need for fire, may be a little hasty in his dismissal of the honey hypothesis. A study of chimps in Loango National Park in Gabon found that a little primate tenacity combined with manual dexterity is more than enough to overcome the efforts of bees to protect their honey and grubs.[3] The chimps of Loango are habitual tool-wielding honey hunters, and they need no fire. While some of the nests they raid are those of stingless Meliponine bees, the chimps have no apparent fear of the notoriously aggressive African honeybees (*Apis mellifera*) and readily climb trees to slurp honey from the modified sticks that they dip into the nests. This tells us not only that our ancestors had little difficulty getting large quantities of honey but that selection for toolmaking and tool-modifying ability was directly linked to a high-carb, honey-rich diet. The chimps' hairy bodies might well make it harder for bees to deliver their suicidal stings, but this doesn't negate Crittenden's thesis. After all, human honey hunters of the Kikuyu and Kalenjin peoples of Kenya actually strip off their clothes to raid bees' nests so that bees don't get trapped in their trousers. And Vedda honey hunters of Sri Lanka are said to be "indifferent" to bee stings when they raid nests.[4] In fact, the protective clothing associated with beekeeping is a recent European affectation; even honey hunters of medieval Wales bore swollen faces from countless stings. So it seems that the pain of a bee sting is far outweighed by the pleasure of eating honey, and our hominine ancestors were probably tougher than the average university professor. But even if they were averse to stings, there are means other than fire to pacify bees. Ethnobiologists Thomas Kraft and

Vivek Venkataraman document twenty-one plants and plant extracts used by honey hunters across Africa, Europe, and Asia to pacify bees without recourse to burning.[5] Whether as bruised leaves applied to the body, broken stems placed near a hive, garlands worn about the head, or leaves chewed up and spat on bees, these plants are all effective in quietening bees and making honey hunting easier for a hairless primate.

As for when humans made a shift from honey hunting to beekeeping, the archaeological record does not give much away. In her epic *World History of Beekeeping and Honey Hunting*, Eva Crane finds the first evidence of beekeeping in Egypt in 2400 B.C. There is a bas-relief from the sun temple of Ne-user-re that depicts a seasonal harvest of honey from what appears to be a stack of ceramic hives not unlike those used in modern-day Egypt. A man is depicted blowing smoke into one of the hives, while two others seem to be storing honey in jars. But ancient Egyptians were habitual recorders of things in stone, so while this is the earliest evidence we have, the practice of beekeeping may well have begun earlier and elsewhere.[6] In fact, the move from honey hunting to honey harvesting and beekeeping is pretty seamless, and in light of honey-hunting practices, it's easy to see how it came about independently in various places. This is because bees are homebodies—they attach themselves to particular places and, given the right conditions, they stick around. So humans, given their own tendency toward territoriality, inevitably find bees attached to the places with which they identify. Out-group humans are prevented from accessing the nests, and so ownership becomes a possibility. The next step is for individuals within groups to claim ownership of particular nests. There is a lot of literature on this practice, which usually involves a honey hunter making an individual mark on a tree and signaling to other humans that the nest belongs to him (honey hunting is mostly done by males).[7] In a small community, the marks are known to all, so someone who finds a mark on a tree can link a particular nest to a particular person. Once ownership of a nest is established in a man's mind, it is a small step to separate the bees from some of the wider ecological processes in which they are otherwise enmeshed. In Djibouti, for example, a man who claims ownership of a nest will encircle the tree with thorns to exclude ratels (honey badgers). The same method is employed in Sumatra to fend off Malayan sun bears. In Poland, a nest might be protected from brown bears with a heavy deadfall made of stone or wood.[8] When a bear intent on raiding the nest climbs up the tree, he trips a hemp rope, bringing the deadfall down on his unsuspecting head. If the deadfall fails to deter bears, then iron spikes and hooks encircling the tree or the nest entrance provide an effective alternative. Other kinds of protection include stopping up holes that might permit entry by pests, insulating nests with straw, and, as with the Cheremis Tatars of northern Europe, saying prayers and making supplications for the well-being of the bees.[9]

In addition to modifying trees and logs to keep predators out, bee tenders make access easier for themselves by attaching doors to tree-bound nests. This is a progression from the act of stopping up enlarged holes in tree trunks through which nests have been raided, in the hope that a smaller hole might attract a new colony of bees. Fashioning a door to serve the same purpose makes a lot of sense. By this stage, the concept of ownership and care is well established in a bee-tender's mind, so it's not a great leap to cut the nest out of the tree and take it home, to be tended there. But there are other ways to arrive at domestication. In some cases, bees make nests in the stone or mud walls of people's houses; once these are made accessible to the homeowner, he need never make another trip to the forest. In other cases, prospective beekeepers deliberately make places at home that are attractive to bees and wait for a hive to establish itself. This is what people do in my wife's homeland of West Shewa, Ethiopia. While some honey is gathered opportunistically from stingless bees who nest in the ground or in tree hollows, the honey of *Apis mellifera* is harvested in a more systematic way. The first step is to construct a suitable home for the bees. Strips of bamboo are cut into half-meter lengths and tied together to create a cylinder similar to a hollow log. The ends are stopped up with wattle and daub, one end being completely blocked and the other with a slot cut into it through which the bees can enter and exit; then the thing is covered with *ensete* leaves to protect it from rain. Having constructed his hive, the farmer gathers up leaves of the wild olive known locally as *ejersa* (*Olea africana*). He makes a pyre of the leaves, above which the hive is hung so that it is fumigated to keep out wax moths.[10] Sometimes the farmer returns to find that the smoldering leaves have reignited and set fire to his new hive, but under normal circumstances the result is a clean, intact hive that is suitable for bees. But how does he persuade bees to occupy the hive? This is effected by means of an attractant: the leaves of the *sombo* tree (*Ekebergia capensis*). These are bruised and placed about the entrance of the hive, and soon enough the bees move in. How do these leaves attract bees? Richard Underhill, a beekeeping blogger from Arkansas, gives an account of his visit to Bonga, a district southwest of West Shewa, where he witnessed the preparation of a hive for wild bees.[11] The beekeepers there use a plant called *limich* to attract the bees. When Underhill smelled the bruised *limich* leaves, he immediately recognized the scent as that of the Nasanov pheromone, a chemical compound that is issued from the Nasanov gland on a bee's abdomen (named for the Russian anatomist who discovered it). Bees lift up their bottoms and fan their wings over the gland to disperse the pheromone, which helps foragers orient themselves toward the hive. It's also used by scouts to mark the location of food and water sources and by workers while queens are making their mating flights. Having no experience of the smell, I can't say for sure, but it's a pretty safe bet that *sombo* and *limich* attract bees by the same means: imitating the pheromone that guides bees to a hive. So here is a first step toward domestication. It begins with a human intervention in the relationship between bees and wax moths and is followed by the use of attractant to draw bees from the forest and into a hive that protects—and separates—them from thieves, honey badgers, and the elements.

Once a hive is established and occupied, it's easy to see how someone might fiddle with it to make it more productive. The obvious place to start is with the hive's design. Eva Crane describes a historical shift in hive design by which "inherited tradition and extemporary improvements" gave way to design by "thought and reason."[12] What she calls rational beekeeping emerged in Europe in the 1600s in parallel with the scientific revolution. In terms of beehives, this was a period of experimentation and innovation in which new materials and manufacturing techniques were applied to making hives that were more productive and easier to use. A plethora of new hive designs emerged, ranging from vertical octagons with removable sections to horizontal, sideboard-style structures with vertical sections and inverted bell jars that could be removed and harvested once the drones had filled them with honey. The jars sat above holes that were made small so that the queen couldn't get inside and lay eggs in the honeyed cells. The overall trend was toward movable frames held vertically and spaced so that bees could move between them and construct cells vertically within the frames. This design essentially copied what bees do in the wild—they build spaced vertical combs—and made things as practicable for humans as possible. In 1852 the Reverend L. L. Langstroth, a Congregational minister in Philadelphia, patented his hive design, which is the predominant design today. Among the features listed in the patent are movable frames, a divider that allows for the addition of supers (additional boxes that increase the number of available honey frames), a separator for queen and brood, and a "trap for excluding moths and catching worms."[13] The Langstroth hive arrived at a time when manufacturing techniques made it relatively cheap to produce, and it has since had a massive impact on honey production worldwide. In parallel with this development, the late modern period has seen the increasing commoditization of bees, including the packaging and worldwide transportation of queens, the scalability and movability of apiaries, and altered selective pressures on *Apis* bees and consequent genetic changes.[14]

Australia had no bees of the genus *Apis* until European colonization, but, as I will show, beekeeping of a certain form was and is still practiced by Aboriginal people, although beekeeping in the traditional Aboriginal sense is quite different from domestication.[15] It is beekeeping of a holistic nature that usually results in the destruction of nests and their inhabitants, but tangentially, ecologically, it is highly beneficial. The bees in Australia are quite different from the aggressive bees that honey collectors in Africa risk life and limb to rob. They are collectively known as "stingless bees." While they can indeed bite and leave a little red mark, they are otherwise quite harmless—unless you're another stingless bee. I'll explore that below.

Unsurprisingly, stingless bees and honey are important to Aboriginal Australians who live within their ranges. They are intricately woven into diets, stories, rituals, and worldviews; there are even bee and honey clans and moieties that largely determine whom people can marry and which foods they can eat. For example, in his work among the Wik-Mungkan of western Cape York, David McKnight found that anyone could collect honey, but there were restrictions on whom they could give it to. These restrictions were primarily due to the shapes of the nest entrances of various bee species. The species that they called *mai kunyan* (probably *Austroplebeia australis*) builds a long, narrow protruding entrance, the phallic shape of which was not lost on the Wik-Mungkan. They called this protrusion the *kunch*, which is also their word for penis. Thus honey from *mai kunyan* bees came loaded with sexual significance that translated into a lot of taboos. While a father could give such honey to his son, the reverse would be akin to a son's giving his father his own genitals and so was strictly forbidden. Surprisingly, it was also forbidden for a daughter to give *mai kunyan* honey to her father, so perhaps the *kunch* represents nonspecific genitalia. But a daughter *was* permitted to give her father honey from *mai atta* bees, who construct a longitudinal nest entrance that sits flush with a tree's exterior and looks like a vagina. And a father could give his daughter *mai kunyan* honey with no concerns about incest. Perhaps the desire to grow strong healthy kids overrides fears of incest attached to the realm of symbolism.[16]

Among the people of the Northern Kimberley—the Ngarinyin, Worora, and Wunambal—Kim Akerman describes a host of social obligations associated with honey. For a young man wanting to marry a particular girl, honey was among the gifts that he was required to give his future parents-in-law. Moreover, the honey had to be collected by the young man's mother, which presumably gave her a potential veto over her son's choice of wife. Honey was also important in group get-togethers. It was "bad form" for a group of visitors to arrive without a load of honey, and in fact even a drop of the stuff was better than none at all.[17] And as with the Wik-Mungkan, there were many restrictions on who could give honey to whom. A young girl could not eat the honey given by a young man to his mother, grandmother, or prospective in-laws; only senior men could eat honey that was presented at rituals; young male initiates' female relatives had to collect honey for presentation during a revelation of sacred objects; and pregnant and nursing women were forbidden from eating honey from the *mindi-mindi* tree. This last prohibition stemmed not from any supernatural or representational reason but from the understanding that the honey, like the fruit of the tree, would cause abscesses of the nipples and prevent the flow of milk. In the same vein, honey from Bauhinia trees was only for consumption by men.

The Ngarinyin people had four categories of tools specifically created for honey extraction. The first was for examination of a hive and could be a rock for sounding out a tree limb, a probe for measuring a hive's dimensions, or a stone hatchet for opening up "peepholes," again to estimate a hive's dimensions. The second was a set of tools for opening the hive: digging sticks, hammer-stones, and stone hatchets. The hatchets were quarried by men, ground and finished by women, and used primarily for extracting honey, although people said that they also used them to knock small mammals from trees. No point in letting a stone hatchet sit idle when there are possums to be had. Honey extraction was often done with the fingers, but people also used wads of fiber to catch spilled honey and a "honey mop"—a sapling with a frayed end that was thrust into a hive to soak up honey. The fourth class of tools consisted of containers; the honey was stored in a watertight bucket made from bark. Ironically, the bucket was sealed with propolis—the wax/resin substance produced by the bees—which was squeezed and chewed to make it malleable. Propolis was also mixed with charcoal or ochre for various other uses. Being malleable and sticky, it was ideal for hafting and repairing stone tools, attaching ornaments, making resin figurines, and making rock art, either as a protective ledge over a painting to prevent water damage or incorporated into the art itself. Propolis is ideal for use with rock art, as it retains a degree of plasticity, whereas tree resins tend to dry out, crack, crumble, and fall off. David Welch describes some examples of rock art in the Kimberley that incorporate propolis, ranging from human figures, to animals, to spirit figures and animal tracks.[18] The use of propolis in art is probably not just a matter of utility, given bees' importance in social and economic relations. Hence a blob of propolis added to a human figure, or even a propolis ledge to protect the painting, might have a great deal of significance to a Ngarinyin person.

In search of a better perspective on Aboriginal Australians' conceptions of bees and honey, I'm at an unlikely place: the Australian National University in Canberra, where I'm catching up with my friend and colleague Natasha Fijn. This is certainly not stingless bee country; it is chilly even in early March, and in four months' time the temperatures will be dipping below zero. But outside the campus café among the sycamores, there is much that my friend can teach me about stingless bees and their engagement with Aboriginal people. Natasha is best known for her extensive fieldwork among the herders of the Khangai Mountains of Mongolia. There, she developed ideas that challenge traditional views of animal domestication, finding instead what she calls a co-domestic relationship of mutual reliance among herders and their livestock.[19] More recently, she's been doing fieldwork among the Yolngu people of northeastern Arnhem Land, where stingless bees are to be found and people are closely connected to those tiny insects. Natasha not only writes up her results but creates films of her ethnographic work, among them a short film called *Sugarbag Dreaming*, in which she follows a family as they search for and open up bees' nests.[20] While the Australian bush is visually modest in comparison to the spectacular mountains and steppes of Mongolia, the Yolngu people in the film transform otherwise mundane scenery into vivid subject matter. In the opening sequence we see the extended family—grandparents, parents, kids, cousins. They are walking among the scrub forest, hands idly perched on hips, tapping on trees and looking up into the high branches, searching for bees. The forest is unimpressive. There are no trees of great height, just stunted, twisted trunks scattered here and there. There is no dappled shade, no leafy forest floor, just a mess of brown grass, occasional ferns, and the bark and leaf litter that eucalypts carelessly leave lying around. But the way the family behaves in their search reminds me for all the world of my own family looking for the Christmas lights in December when it's time to decorate the tree. And that is the power of the opening sequence of *Sugarbag Dreaming*: we find a family searching for bees, not out there in the bush but very much at home.

Soon enough they find a near-dead tree that is occupied by bees. Gazing up at the thing, Natasha is told that there is a wax entrance and the bees are flying in and out. She is incredulous. "So how did you see that?" The answer: they saw the bees. Perhaps unsurprisingly, the locals have a better eye for tiny bees than an anthropologist. Apparently, they also find them by smell and by the barely audible low humming sound.

Once the nest is found, a young woman makes short work of felling the tree with an axe. Then, while the children wait expectantly with their plastic containers, Grandma takes the axe and begins cutting notches into the hollow trunk. She is trying to determine the extent of the nest. While she hacks away, we are given a lesson from another elder. He tells us the Yolngu names for the larvae, honey, and yellow waste matter that bees drop from their nest entrance. He tells us how the bees favor two particular tree species: stringybark and *djarmal*. Practical stuff. He tells us which species is associated with which clan. The bee with the "long nose" (the tubular nest entrance) is Gulumulu clan within the Dhuwa moiety, the short nose Marakulu clan. This is all elementary, just the entrance to a nest of deeper meanings. Even the "nose" descriptor is probably the PG version of a more complex account that we can only guess at. According to Natasha, knowledge is power among the Yolngu, and nobody is going to give too much away here in the forest during an interview; this stuff is learned through ceremony. Later in the film, we hear from Yirralka ranger Yumitjin Wunungmirra (Jimmy) that the family who stands for Yolngu laws and culture acquires these laws from the sugarbag complex. In other words, bees and human associations with them form the basis for Yolngu laws and, by extension, their ceremonies, song lines, stories, and paintings. What's more, Jimmy has a right to say this, while others may not, because Jimmy is of the Dhuwa moiety. Indicating the bees' nest, he tells us that Dhuwa is his mother, but this is not just an allusion to his mother's moiety, nor is it metaphor. In terms of cosmology, the sugarbag complex—the honey that his family are extracting in the background, the half-dead tree that they just felled, the bees and larvae, and the propolis—is not just his mother's totem; it is literally his mother. This is crucial to the distinction between Western and Aboriginal relations with bees: for the Yolngu, the human figure of the ranger's mother and the other elements are all part of the one being and cannot be treated in isolation of one another.

Cut back to the tree, and there's pandemonium. The bees are in emergency mode while the women demolish the contents of the nest. As it turns out, the bees occupied a hollow channel inside the trunk, about 1.5 meters long and some five centimeters in diameter, that's been neatly exposed by the head of an axe. Children's fingers scoop up the contents of the hollow—honey, propolis, pollen, larvae, bees, everything. They stuff some into their mouths and drip the rest into their plastic containers. Natasha's filmmaking style brings us right up to the action, and I can almost taste the sour, aromatic honey and feel the larvae getting stuck between my teeth. People are chattering and laughing and slurping sugarbag off their fingers or dripping it into containers from upturned lengths of tree. It's a celebration.

Sporting a summer dress and a china-doll hairstyle, Natasha doesn't look like someone who can withstand the extremes of the Mongolian steppe or Australia's tropical north. But she is driven and inquisitive and her knowledge is expansive. She explains to me that for the Yolngu, "sugarbag" is a term used not just for the honey but for the entire cultural and cosmological complex surrounding stingless bees. "They're not thinking of that stingless bee as an entity unto itself. They're not just thinking, 'Here is a stingless bee, and we're gonna get honey from it.' They talk about it as sugarbag. That encompasses the actual nest and the wax and the larvae and the bee and the stringybark tree that it's in and then the time of year and the whole surrounding ecosystem." This is how a ranger can talk about sugarbag as a basis for laws, ceremonies, song lines, and paintings. When Yolngu see connections between bees and wider ecological processes, they see themselves infused into these connections. This is why I suggested that the bees are "kept" in a certain fashion: through ceremony, ritual, and tradition, the Yolngu maintain these connections that ensure the integrity of the whole system of which they and the bees are integral parts. And the bush wherein they "keep" their bees is not "the wild"—it *is* the *domus*; it is home.[21] Ironically, it is the importance of these ecological connections that, according to Natasha, largely precludes the possibility of Western-style domestication of stingless bees. "They were never on this road to domestication, because it just lies outside the knowledge structure that they already had, that they thought worked perfectly fine anyway. It's like, why would you? It's counter to the philosophy of how you engaged with the land. It would be only if they were adopting the philosophy of the colonial wider Australian perspective that they would do that."

Not surprisingly, the wider (and whiter) Australian perspective is pressing heavily, albeit unwittingly, upon the traditional Yolngu philosophy. With all the good intentions and misunderstandings of zealous missionaries, government organizations think that they can save Aboriginal people from themselves by incorporating a bit of nativeness into a capitalist paradigm. Hence the attempts to get Aboriginal people to engage in stingless beekeeping so they might derive income from their efforts. Natasha explains, "There are programs, like with ranger programs where, I think, larger Australia—managers of the programs all think, 'Oh, it'd be much easier if we brought in hives and just put them in the community and made the products.' Priorities from the wider Australian framework are saying, 'Here, this is what we think you could do to make this productive, and you could be part of the economy.'"[22] This betrays a perspective that misrepresents traditional economies; it assumes that finding food in the bush is difficult. It reiterates a nature/culture dualism that opposes humanity and nature, seeing the bush as dangerous, precarious, and resource-scarce. But for Yolngu, living with the bush is not precarious and neither is finding sugarbag; these are practices of the home. "For Yolngu communities," Natasha explains, "it's easy enough to just go out and get the honey anyway. For them, it [keeping bees in hive boxes] wouldn't be making life easier for themselves because they know exactly where the location is of their particular nest or nests in their area. It's not an onerous task. It's an enjoyable task." These people are not stupid. Faced with the choice between harvesting the fruit of the forest or entry into a capitalist edifice on the ground floor, they take the sane option. Meanwhile, the rest of us have to answer for the poor choices of our ancestors.

The Euro-Australian way of engaging with stingless bees is as far from Yolngu ontologies as Southampton is from eastern Arnhem Land, and keeping bees in backyards that stand in opposition to the bush fits seamlessly into the emergent logics that are woven into the fabric of wider Australian society. Stingless bees are loaded with potential for an Australia that sees ecologies as conglomerations that can be unmade, broken down into components that in turn can be valued—and exploited—separately. Stingless bees are productive; they take nectar and transform it into honey; they collect tree resins and mix them with wax to create propolis; they pollinate crops and orchards. Stingless bees are low maintenance; once a hive is established and the internal structures are set up by the bees, it can pretty much be left alone until harvest time (although this is not always what people do). Stingless bees can be reproduced: when a hive reaches maximum weight, it can be split and voila: one hive becomes two. Finally, stingless bees are native; they have the cachet associated with species endemic to the Australian continent. In the November 1997 issue of *Aussie Bee*, the bulletin of the Australian Native Bee Research Centre, there is a letter to the editor that describes the writer's bees as preferring Vegemite to jam, peanut paste, and honey. The editor responds that the bees are probably attracted to the salt, but "with a taste for Vegemite, they're genuine Aussie bees, aren't they!"[23] With a combination of native cachet, utility, and environmental cred, hives of stingless bees are being bought, sold, and shipped across the country to the farms and backyards of Australians who subscribe to an ontology that until two hundred or so years ago was not of this land.[24]

But the unmaking of stingless bees is not a conspiracy, nor is it a project of profit-driven multinationals. It's a grassroots movement driven by enthusiasts and bee lovers in suburban backyards where profit takes a backseat to passion. It fosters networks of individuals who tinker with beekeeping and share information about these captivating little insects. One of the most prominent human figures in the stingless bee industry is Tim Heard. Tim began his attachment to stingless bees as a pollination biologist, but his passion for bees outweighed his passion for academia, so he now works full-time with his business, Sugarbag, promoting and selling stingless bees and related services and products. In fact, Tim wrote and published the most comprehensive and informative book on stingless bees that I have seen: *The Australian Native Bee Book*. Amid a plethora of exquisite photos that celebrate the macro lens as much as they celebrate bees, Heard takes readers from bee phylogeny and evolution, through bee behavior, to what almost seems a logical consequence of all this knowledge: keeping bees in boxes. In one part of the book there is a distinction between two genera of stingless bees based on microscopic differences between hair density on their hind legs, while in another section, species of the same two genera are distinguished by the volume of hive required for their keeping. The book is written with all the academic rigor of a scientist yet at the same time all the empathic charm of a human being. After describing age polyethism in bees—how, toward the end of their lives, hive workers graduate from the warm seclusion and security of the hive to the rigors and challenges of foraging out in the world—Heard says, "Spare a thought for the astonishing hormonal, neurological and behavioral changes that each worker must experience to allow her to play both roles (hive worker and forager) during her life."[25] This kind of empathic writing is anathema to the scientific community in which Heard spent many years, so it raises questions, at least for me, about what kind of ontological shift follows the transition from scientist to beekeeper.

Tim is also unreserved about the idea that stingless bees are being domesticated, although he doesn't explore this subject at length in the book. In a half-page section titled "Breeding Better Bees," Heard points to breeding and hybridization programs with honeybees to suggest that the same might happen with stingless bees. But, he says, before any breeding program takes place, "we need to clearly define its goals."[26] Increased honey production seems like an obvious goal, given that honey is the primary product of honeybees. But stingless bees are only one fiftieth the weight of honeybees, and a viable stingless bee hive produces only one kilogram of honey per year. Even with intensive selection for honey production, would it be possible to compete with honeybees for a slice of the honey production pie? At face value it seems unlikely, but then stingless bees are more resistant to the pathogens that are causing problems for the honeybee industry; they are not affected directly by colony collapse disorder (CCD) or varroa mites.[27] So in a world facing up to the decline of honeybees, stingless bees might prove their worth to humans, not by being superproductive but simply by not being extinct. Stingless bees' honey might also be valuable for its quality rather than its quantity, in that it has therapeutic potential. Regardless of the flowers on which the bees feed, stingless bee honey seems to have excellent antimicrobial properties. One reason for this may be that it contains flavonoids that are infused into the honey from the tree resins in the propolis pots in which the honey is stored.[28] So it may be propolis production, not honey production, that guides selection of stingless bees. In fact, there is a tube of propolis toothpaste sitting in my bathroom, which, although it doesn't say how much propolis extract it contains, suggests that there is a market for the substance if it's associated with health. Another reason why stingless bee honey might have antimicrobial properties is that it contains beneficial microorganisms that have evolved to thrive in the high-acid, sugary honey and work to preserve it. These are the sorts of connections that might better align science with a Yolŋu ontology.

Of course, selecting for increased production, or for honey or propolis quality, or for disease resistance will be a more or less systematic process akin to breeding faster-growing chickens. But the simple act of keeping stingless bees in wooden boxes in suburban backyards is a form of unmaking, altering all sorts of selective pressures on these insects. It is these peripheral forms of selection that I'm interested in, so I'm paying a visit to Tim Heard at his home in a hip suburb of Brisbane. The house is unremarkable, a weatherboard home sitting on stilts to accommodate the sporadic rise of the Brisbane River. The only clue that it's a nexus of the stingless bee industry is the unusually large number of tiny black bees buzzing around in the garden. Tim greets me warmly at the upstairs entrance, well above high-water mark. He is handsome and athletic for a fifty-something and very welcoming. But he's also very busy with his bee business. He enlists my help loading some equipment into the car and takes me with him on a drive across town to do a hive split. Our first stop is his brother's house, where hive-box production is humming along. It is yet another suburban home, unusual only in the number of hives in the yard. On the way around to the back of the house, Tim stops me at one of the hives on the garden path and indicates a large lozenge-shaped insect with translucent brown legs standing patiently on the side of the hive. "That's an assassin bug. They hang around the hive entrance and prey on bees. It'll inject saliva into a bee, liquefy the insides, and suck them out." This sounds like a script for a sci-fi movie. I'm expecting Tim to swat the assassin bug, because this is the remedy he suggests in his book. To my surprise, he moves on, leaving the bug sitting on the hive. I suppose that if a hive contains thousands of bees, an assassin bug would be hard pressed to endanger it by liquefying one bee at a time.

In the garage, Tim's brother is assembling hive boxes. It's certainly not an exploitive mass-production facility, but there are enough hive boxes to indicate that this cottage industry is booming. Tim collects some equipment and the parts for a hive box, and we drive to a leafy suburb where a renovated Queenslander-style home sits on about half an acre of suburban garden. We're greeted on the veranda by Brian and Margaret, a couple of professionals on the verge of retirement. Margaret takes us to a corner of the veranda, where a hive sits on a little platform that has been attached to one of the posts. Tim remarks that the hive is in an excellent position: it's nicely shaded and sheltered from southerly winds. These are well-to-do bees. He moves the hive to a low table on the veranda where he'll do the split. Taking his hive tool, a cross between a paint scraper and a crowbar, he prizes open the hive and exposes its occupants to the elements for the first time in their lives. Tim tells me that the bees are *Tetragonula hockingsi*. These bees are different from *T. carbonaria* in that they build amorphous brood structures, as opposed to the spiral brood structures of the latter species. He indicates one bee with a large striped abdomen—the queen. She moves quickly across the brood and disappears into the web of propolis while the other bees bustle about. Presumably they're in damage-control mode, given that their hive has just been rent and opened to the elements. Tim indicates a white brood cell on the periphery and tells me this is a queen cell. These cells produce virgin queens who can live in the hive. Unlike honeybees, they can stay there indefinitely and will not be killed by the reigning queen. If the queen dies or is failing, the workers select a virgin queen from among the residents and allow her to go on a mating flight. She returns to the hive and brood production continues. The previous queen, whether dead or still kicking, will be taken and dumped unceremoniously outside the hive. In his book, Tim compares stingless bees to a democracy in that the workers choose their queen. Interestingly, in *T. carbonaria*, if a queen dies and no virgin queens are present in the hive, a larva can move into an "emergency" queen cell—a larger brood cell that is provisioned with food—and thereafter develop as a queen. So it's a democracy with a degree of social mobility.

As Tim cuts through the brood, the bees become defensive. I notice bees crawling through my hair and feel them biting the back of my neck. It's a pretty feeble defense, though, which suggests that these bees did not evolve under a particularly high threat of predation—at least not until people arrived in Australia some sixty-five thousand years ago. Tim takes a new bottom section that we brought along and attaches it to the occupied top section. Then he adds a new top to the occupied bottom section, and where there was one there are now two hives. He seals the joins with duct tape and explains that the bees will need some time to repair the internal structures and get everything back to normal. The section without a queen will also need to send a virgin queen out on a mating flight. Tim sticks a clear acetate sheet over the top of one of the hives. This hive is a present for Brian's grandson, who will want to monitor it as it develops. Tim then returns the other hive to the platform attached to the veranda post. The bees crowd around the front of the hive, exhibiting a defensive reaction, but once that settles down it will be business as usual and see you next October for another split. This is bee reproduction suburban-style.

The following day, we're back at Tim's house on the rear veranda looking out over the garden and talking about the future of stingless bees. There are a couple of researchers working at Tim's house: Toby, an ecologist, and

Lauren, a grad student. They're both busy tapping away on their laptops. Tim is sipping tea and very engaged in answering my questions. I ask about differential selection, the way in which different selective agents might affect the reproduction of kept bees in comparison to their wild cousins. Toby chimes in and I'm taken aback by his level of certainty. "Oh, absolutely! We're reproducing hives by splitting, and this is based on weight. When a hives gets to a certain weight, we split it. So the hives that put on weight the fastest are the ones that are being reproduced more often." At least that's where things stand at present because, as Tim tells me, it isn't honey or propolis that is the most profitable product of stingless bees; it is the hives themselves. This means that rapid weight gain and the ability to withstand frequent splitting are the traits that are being selected for, thanks to the current demand for hives. If for some reason honey should become more profitable than hives, then rapid weight gain will still be desirable, but hive splitting will be done less frequently because it inhibits honey production. And this in turn will depend on other factors, such as the market for honey and propolis. It will also depend on the demand for hives for crop pollination and enjoyment and on the contemporaneous persistence of honeybees in the face of CCD and other threats.

Tim expands on the difference between wild and suburban reproduction. "The other thing we've provided them with is more *opportunities* for reproduction," he says. "This is because reproduction for these stingless bees is quite risky in the wild. Why? Because they have to find a new suitable nest site to reproduce. They have to find that within their flight range." Stingless bees are not like honeybees, who simply form a swarm and go looking for another nesting site. Stingless bees need to find one in advance and prepare it before they move in. They normally fly only five hundred meters or so from the hive, so their choices are quite limited. "We've taken away that constraint as well," Tim goes on. "And even if you're splitting based on a particular time period as opposed to weight, I think you've got selection, because if you're going to split a weak hive as opposed to a strong hive, the chances of the weak one not surviving the split and one half or both halves dying are greater. We're still selecting."

Okay, then what about things like reduced flight range or robusticity? Do you see any selective differential in urban bees? "It's possible," Tim responds. "Let's break those two things down. I think the most significant point you make there is the robusticity or the toughness, the . . ." Independence? "Yes, the independence! Basically, the ability of these bees to survive in the wild."

Without human intervention. "Yeah, without intervention. Are we affecting that?" I give the extreme example of domesticated corn that can't even reproduce without human intervention. "Yeah, the husk, which was advantageous to domestication because it allowed humans to store that product a bit more easily, became a disadvantage to the plant, naturally, because the seeds can't disperse. So what are we selecting for? To some extent, we're protecting them from natural enemies. Some beekeepers use a flyswatter to wipe out some of the pests that come close to the hive. Like phorid fly. And a lot of people are using traps. Therefore, if they were diverting resources away from defense and putting it into producing more food, the hive would be putting on more weight, and it would be split more quickly. That would be changing the genetic propensity of the hive from self-defense to increased foraging. It's possible."

What about keeping bees in an urban environment? I ask. It's a pretty easy place to find food compared to the forest. "Yeah. In the wild, they really need to be using their abilities to make sure that they're harvesting in the most efficient way," Tim replies. The resources are far more scattered than in an urban environment. "There's absolutely no doubt about that. Most beekeepers harvest bees from the wild when they start their beekeeping enterprises, but then they soon realize that if they can keep them in urban environments, the bees do a lot better. Out in the bush, it's tough. The bees have to use every ounce of skill to the limits of their small brains to exploit their environment in an efficient way. Whereas in urban areas they just need to leave the hive and head to the nearest floral scent and they can harvest an abundant amount of food."

As Tim talks, I'm looking out over the garden below. It's October in Brisbane, springtime; the air is warm and the flowers are putting on a show. I can see at least four hives in the garden, and tiny black bees are out and about collecting pollen, nectar, and tree resins the likes of which their bush cousins can only dream of. But what of this practice of keeping hives close together, as Tim does in his backyard? I know from his book that drift fighting can occur when bees arrive home at the wrong hive, and these fights lead to lost production. In fact, Tim recommends painting different symbols on closely spaced hives so that foragers make fewer mistakes. I ask him about drift fighting, and he explains what he calls fighting swarms, which form when the bees from one colony invade another colony with the aim of taking it over. "What they do when they get in is a little bit unclear," he says, "but we think they kill the old queen and then install their own queen. What that means is that, as a species, they're highly defensive against that happening." They mount these aerial, defensive displays even when they only think they're under attack, Tim explains. "If you move them into high densities in the orchards for pollination, and the bees come back to the wrong colony, they're not attacking another colony but they're drifting. The bees in that colony think they're being attacked, so they put on an aerial display to show the invaders, 'we are many and it's not worth attacking us.' I can show you that right now," Tim offers. "If I take two hives and swap them, then the foragers will be coming home to the wrong hives and you'll see a reaction. When they go into that defensive mode, they stop doing everything else. You want to minimize that." I ask him, how do you minimize it? "Well, you need to stop bees going to the wrong hive." Tim explains that if you look at a forest from above, it looks very homogeneous—full of texture, but homogeneous, with no real discernible patterns. Yet bees are very good at learning visual patterns that enable them to find their way home. Researchers have proved that they can learn to recognize shapes. I ask him whether, in terms of selection, there would be any difference between wild and domesticated bees in terms of the demands on bees' cognitive capacities, given that backyard beekeeping provides a much less homogeneous environment than bees in the wild are used to. "That's true," he says. "We published a paper on this recently showing that their homing success was greater in open areas compared to forests, which is a surprise." So how much easier are we making things for suburban bees by providing them with less homogeneous surroundings? It would seem that we're making it easier for them to home, but how much is that really going to affect their cognition?

We leave the question hanging while Tim takes me downstairs to the garden, where he intends to show me a defensive reaction in his bees. It's a typical rambling suburban garden, with unkempt lawn, some shade trees, tomatoes ripening in a veggie patch and beans growing up trellises on the fence, the sort of place that accumulates childhood memories and faded footballs. There are hives here and there around the garden, their occupants active in the warm October sun. Tim looks about and points to a hive in the shade of a brush box. "We'll swap that one with the one over here. That should get a result pretty quickly." He picks up a hive from a ceramic pedestal beside the veggie patch and carries it over to the tree, then returns to the veggie patch with the other hive and places that on the pedestal. We sit on our haunches and watch as the first forager returns to what is now a foreign hive. She makes for the entrance but is set upon by one of the guards; they lock mandibles. I look at the step in front of the hive and another two bees are locked in mortal combat. They fall to the ground, where they will surely die biting each other, and a wave of guilt sweeps over me at our deception. The bees inside the hive are emerging; soon there will be a defensive swarm in front of the hive and a host of unsuspecting foragers returning to a very unpleasant welcome. I tell Tim that I can see clearly what's happening and he's more than happy to wind up the experiment. He quickly returns the two hives to their original positions. There are already the beginnings of a defensive reaction in the second hive as well, but Tim assures me that things will settle down pretty quickly. I thank Tim for his generosity and we say goodbye. The bites on the back of my neck from the previous day have faded, but in time I will find that stingless bees bite more deeply than I suspected.

Back at home on the north coast of New South Wales, I'm catching up with a friend who keeps stingless bees. Tara manages the local health food store and, like a vast number of keepers of stingless bees, she keeps them just for pleasure. Her relationship with bees began when her dad made her a hive box. It's an exquisite little construction with a roof and base reminiscent of a Japanese shrine. It also has a viewing screen at the top, so we can see the goings-on inside. The only drawback of the gift was that it had no bees. Tara's dad lives in Canberra, so he wasn't able to stock it before giving it to her for Christmas. So Tara went online and managed to source a colony of *T. carbonaria* from a guy in Brisbane who promotes and publishes a blog about stingless bees. She transferred the colony into the hive box and put it on top of an old washing basket stand that she bought at a thrift shop. She weighed everything down with a brick and placed it under a large cadaghi tree in her backyard.

At the time, Tara didn't recognize the species of tree that shaded her hive, but she soon learned about the relationship between cadaghi trees (*Corymbia torelliana*) and stingless bees. These trees are native to the rainforests of North Queensland. They have bright green leaves and bristly, cream-colored flowers, and the gum nuts they produce (which are not actually nuts but seed capsules) are shaped like little Greek urns. While these trees are native, the Australian government considers them noxious weeds. They grow to tremendous heights—up to thirty meters—and their dense canopies tend to shade out understory species, altering the diversity and structure of the forests wherever they're introduced. How do they disperse their seeds? Via stingless bees. The sticky resin inside cadaghi seed capsules is irresistible to stingless bees, who crawl inside the capsules and inevitably emerge with a few seeds stuck to their bodies. The seeds are an unwieldy load for a stingless bee flying home, so they do their best to shake them off, facilitating dispersal of the seeds. They often fail to get rid of all of the seeds, so stingless bee hives frequently have a plastering of red cadaghi seeds around the hive entrance and cadaghi saplings growing on the ground below. In this way, stingless bees have been instrumental in the spread of cadaghi trees beyond their historic range and the consequent changes to forest composition. It's quite ironic that a native bee is facilitating the invasiveness of a native tree.

For some time, keepers of stingless bees considered cadaghi a threat to their bees. Some believed that cadaghi resin melted in hot weather, leading to the collapse of nest structures and the blockage of hive entrances; others believed that it released fumes that killed their bees. Such beliefs are symptomatic of the "bees-against-the-world" approach to beekeeping that leads to the sorts of separations that this book is about. The logical consequence of these beliefs is the attempt to remove any cadaghi trees on which the bees might be feeding. Tim Heard is of the opinion that cadaghi resin is not a threat and that the trees and bees have evolved a mutually beneficial relationship. This is because cadaghi resin has been shown to maintain its consistency at temperatures beyond those at which bees have died. So the myth about cadaghi probably arose when beekeepers opened up hives to find the occupants dead and a sticky mess of cadaghi resin flooding the hive. They blamed the resin for the bees' demise, when in fact it was plain old heat that killed the bees and softened the resin. As Tim points out, it's "not in the trees' interest to kill their dispersers."

Having seen that the immediate presence of a cadaghi tree had no ill effect on her bees, Tara is happy to let the bees enjoy the sticky resin (which, incidentally, is like mozzarella cheese in color and texture—though not in taste). She's also skeptical that bees could be harmed by cadaghi and strongly believes that bees are well adapted to conditions in Australia. "They're pretty resilient—they're heaps more resilient than I thought they were at the start. I don't reckon they really feel much danger. My neighbor sprays Roundup, but that's never, ever affected them in any way. That was one I was super concerned about. I reckon they're pretty staunch."

But Tara's view of her bees' resilience was challenged when she set about propagating a second hive. Tim Heard had shown me the splitting method of reproducing hives, dividing a hive in two and cutting through the brood. Tara refers to this method as pretty "gnarly" in that it destroys nest structures and leads to a lot of repair work for the bees. But this is not the only way to propagate hives. There is another that at least in theory is less destructive: the eduction method. This involves attaching an empty hive box to an existing hive by way of a connecting tube attached to the hive entrance. The bees, finding a big empty space between their existing hive and the outside world, set about filling the new hive box with stores of food and brood cells. When the new hive appears to be able to go it alone, the beekeeper makes a hole in the connecting tube, and the bees in the first hive use the nearest exit to the outside world—the hole—while those in the new hive produce a queen who goes on a mating flight and starts producing a brood. After some months, Tara was satisfied that the new hive box had a queen and was producing brood, so she removed the connecting tube and one hive became two.

One month later, I'm visiting Tara at her store and she has the shell-shocked look of someone who's been under siege. It turns out that the eduction method of hive propagation is not as gentle on bees as she'd hoped. Not long after Tara removed the connecting tube between her two hives, she noticed a green tinge covering the food stores and brood: fungus. She didn't know what the consequences of a fungal growth might be, but she wasn't going to wait and find out. Instead, she went online and found a treatment for fungal infections in hives: apple cider vinegar. Fortunately, she had a ready supply of the stuff in her health food store, and organic vinegar at that. She put some in a spray bottle and sprayed a 100 percent solution directly into the hive. This seemed to kill off the fungus, but that was only the start of her problems. The next time she looked through the viewing pane she noticed some grubs in among the internal structures. A hive beetle had somehow gotten past the guards lining the entrance tube and entered the hive, whereupon she laid her eggs. In a "strong" hive, the bees would normally keep hive beetles out, or if one gained entry they would quickly locate and mummify the creature alive by coating her in resin.[29] But in Tara's case, the hive was newly established and weak. The bees failed to disable the invading hive beetle, and her newly hatched larvae, if left unchecked, would demolish the insides of the hive and starve its residents. Tara saw no option but to intervene. One by one she picked the larvae from the hive and squished them between her fingers. But as soon as Tara had the infestation in check, another population of larvae appeared in the hive. This time it was syrphid fly. These flies look remarkably similar to wasps and are often found lurking around hives. In fact, syrphid flies are known to appear within minutes of a hive's being opened. Given the chance, they will lay their eggs directly inside a hive, but since open hives are rare, syrphid flies normally lay their eggs on the outside of hive boxes near gaps or joins. The larvae hatch and enter the hive through these gaps. If the bees have not sealed the hive properly—presumably the case with Tara's hive—then the larvae gain entry and grow fat on the honey and pollen stores within. They utterly destroy the hive before growing into adult syrphid flies and setting off in search of another hive. Again it was left to Tara to remove each larva by hand and deal them death by squishing.

According to Natasha Fijn, Yolngu people see the predators of stingless bees in a completely different light. For them, the "bee-flies" are "restorers of life" for the bees. In fact, they are seen as being just as important as the bees in terms of pollination of the stringybark trees and just as crucial to the integrity of the sugarbag complex. This is why the sugarbag ceremony includes the dhumarr—a large didgeridoo. The dhumarr represents the drone of the bee-fly hovering around the nest. Predators such as syrphid flies have a place within the sugarbag complex. Not just the bees but the entire system, including predators, is subject to the care of the people. This is a crucial distinction between the Yolngu philosophy and wider Australia, and it marks a critical difference between Western domestication and Indigenous ways of relating to stingless bees. For the Yolngu, the system itself is the subject of care. If something threatens the system, that is cause for concern, but there can be no crisis when one part of the system threatens another. That is something that the system will sort out for itself and everything will be okay, as long as they perform the correct ceremonies and observe the relevant timings. The wider ecology is that which is kept intact, and honey from stingless bees is a result of the integrity that they strive to maintain.

Meanwhile, for Tara, the totality of the local ecology consists of the bees, herself, and anonymous plants in the gardens of neighbors with whom she has no relationship. It is Tara and her bees versus the world, and the world includes fungi, hive beetles, and syrphid flies, seen as threats to the system rather than as integral parts. This is what defines the unmaking of stingless bees in Australia: the isolation of hives and beekeepers as nucleated systems within anonymous and often threatening ecologies, the creation of interdependencies between bees and humans, and the severing of connections that entangle bees and humans within complex cosmologies that in turn are inextricable from ecologies. Who would have thought that a relationship with five thousand stingless bees could be so lonely?

Two weeks later, I answer a knock on the door to find a grinning Tim Heard on my doorstep with the beehive that I ordered. I am now a keeper of stingless bees, engaged in the very process that I thought I was critiquing. Tim is on his way down to Sydney, where he will speak at a seminar. He's bumped me up on the list of people waiting for hives, because it's a bit safer and more convenient to deliver a hive in person than to ship it via the post. We march up the hill to the back of my house and I show him the place I have in mind for the hive. I know my yard; I know the path of the sun and where the cold southerlies blast the side of the house. So I've selected a sheltered spot beneath the bedroom window where there is limited summer sun and a clothesline on which I can rig an awning for shade. My plan is to bolt a horizontal post to the wall and mount the hive over an ant cap on the post. But Tim thinks that's unnecessary. It's more important, he suggests, to be able to move the hive forward and into the sun in winter and then back into the shade in summer. So he improvises a pedestal with some bits of timber and sets the hive on that. We talk bees for a couple of hours and Tim leaves me with my charges.

The next morning I wake up early and hurry outside to introduce my bees to their new world. There is a vented stopper in the hive entrance that allows air to circulate but prevents bees from escaping during transport. I remove the stopper and stand back to watch as the first bee emerges. She crawls to the edge of the landing block, turns, and reenters the hive. I can't imagine what that was about. Before long, a second bee appears at the entrance. This one takes flight, and I watch as she turns and hovers about a meter from the hive. This is an orienting flight. As when bushwalking, you turn around and look back to fix an image in your mind of what the scene should look like on your way back. But this is as far as she goes. She flies back to the hive entrance, alights on the landing block, folds her wings, and reenters the hive. Meanwhile, other bees are emerging and taking flight. They all fly to about one meter away before they about-face to get a look at the hive, and then they return. It's like a swarm of real estate agents taking photos for a property listing. But as they fly out and return one by one, I notice that the distance they fly from the hive is increasing. I see bees hovering farther and farther away, until before long they are off into the wilds of suburban New South Wales looking for food and resin to bring home. They seem happy with their new location, and I'm happy about that. It seems that all there is left to do is ensure that the hive doesn't get too much sunlight, and in about ten months I'll be harvesting some sweet honey.

The very next day we receive an unwelcome visitor. I notice a black and yellow wasplike insect hovering about the hive. It's a syrphid fly! I'm truly amazed at how quickly she found the hive in my backyard, but I'm also pretty annoyed that she's paying it so much attention. I have no idea how well established my bees are. What if they haven't sealed up the hive and this wasp-alike lays her eggs around the gaps? I imagine tiny larvae wriggling their way into the hive and fattening themselves on the contents before a swarm of adult syrphids emerges looking for more hives. Tim guarantees his hives for twelve months, but getting my money back wouldn't compensate for the heartbreak of having this particular hive devastated by a larval infestation. I grab a tea towel off the clothesline and swat wildly at the syrphid fly, who is infuriatingly adept at avoiding my swipes. She zigzags off into the garden, leaving me out of breath, wringing the tea towel in my hands, and trying to think of a way to keep this fly away from my bees. I realize that I'm just going to have to trust that my bees have the numbers and resources to shore up their defenses. But I'm also in awe of this black and yellow insect who is so adept at locating the hives of stingless bees. Yolngu people appreciate syrphid flies, because it's much easier to spot one hovering about than it is a tiny black stingless bee. Syrphid flies are like black and yellow flags indicating the presence of a hive.

One week later, I'm not happy with the timber pedestal the hive is sitting on. What if the ants in my garden get word of the honey and raid the hive? They were certainly quick to find the honey jar in the pantry and would have taken it, jar and all, had I not moved it to the fridge. So I go with my original idea of mounting an ant cap to a horizontal post and setting the hive on that. This also attaches the bees to my house in a very concrete way. The wall keeps my hive well above the ground, and the eaves provide shelter from the sun and rain. When winter comes around and the days get shorter and colder, I'll build a dolly with an ant cap so I can move the hive forward into the afternoon sun.

I'm still hung up on the resident syrphid fly, who is constantly hanging around my hive. It's become a morning ritual to go to the backyard and swat at her with a tea towel. She's utterly unfazed by the attention. She buzzes off to a safe distance and within a minute is back at the hive, hovering nearby like a parking officer waiting for a meter to expire. At least I can understand how she found the hive so easily. I notice it exudes a delicious smell, something like a mixture of honey, earth, and sandalwood. It funny, but the smell is quite familiar to me. I've smelled it many times in the bush, but I never realized that it was the smell of a bees' nest. Between the syrphid flies and the powerful aroma, these bees are not very good at disguising their location.

I've been reading about people feeding their bees. Tim Heard's book recommends a 30 percent sugar solution with a bit of rosewater as an attractant. They can also be fed soy protein as a pollen supplement; purists with money can even buy bee pollen itself from the health food store. I convince myself that it makes sense to feed my bees. Winter is around the corner, and that's when they'll be depleting their food stores. There's also our friend the syrphid fly, who is set to breach the hive at the first sign of weakness. So I turn to the internet for a feeder design. There's a dazzling array of choices, but in the end I settle on one created by Nick Powell in Queensland, the same guy who supplied the bees for Tara's hive box. It's a very elegant design, using the neck of a soda bottle and a milk bottle cap. I cut the soda bottle neck down to size, make some holes and slots for the sugar solution to seep through, and plug it into the milk cap. I want to give my bees the best food available, so I pick up some organic sugar from Tara's store and boil it into a solution before adding a drop of rosewater to make it smell nice. Once cool enough, I tip some into the feeder and take it outside to my hive, where I expect a swarm of bees to converge on me and the feeder.

It turns out that bees are not as perceptive as I anticipated. I kneel down in front of my hive and slide the feeder under the gable, behind a bit of four-millimeter mesh that I installed to exclude other, bigger insects. Apart from the usual gaggle of bees hovering behind me when I'm in front of the hive, there is no interest in what's going on. I figure the bees will find the food soon enough, so I sit and watch for some time. I see the usual traffic of unburdened bees departing, and returnees with blobs of pollen on their legs reentering the hive, but not a single bee shows an interest in the plastic container full of food that I've just placed directly above the entrance. After twenty minutes of this, I decide that they need some guidance. I dip my finger in the solution and smear some right outside the entrance. I dip my finger in again and smear a trail from the hive entrance to the top of the hive, directly in front of the feeder. Already some bees are taking notice of the sugar solution at the entrance. Gradually, the bees move higher and higher up the hive, slurping the rose-scented sugary stuff I've provided. Within two hours they are clambering onto the rim of the feeder and having a sip before flying off into the neighborhood. Success.

Within days, news has swept through the hive that there's a free meal to be had right on the doorstep. I watch as each forager emerges from the entrance, crawls up the face of the hive, and takes a sip of sugar solution before launching into the air and heading off on a foraging trip. Within four days the feeder is empty, and I realize that I haven't really given this much thought. My imagination stretched only as far as their first feeding—I was so caught up with feeder design, solution consistency and smell, and teaching the bees to use the feeder, that I didn't consider what would happen when the bees actually finished their food. But there it is—an empty feeder sitting atop the hive. And bees are still visiting it. In one respect, I've created a dependency. Even though I'm sure the hive will flourish should I never feed them again, at some level the bees must be dependent on the sugar solution—thus they keep visiting the empty feeder in expectation. I've also created an obligation. I feel sympathy for the bees with empty tummies; I feel bad that their food has run out. I take the feeder downstairs, top it up with more sugar solution, and return it to the top of the hive, where the bees quickly find it and resume their pre-foraging ritual.

From my position of having done it, I can see much more clearly how the process of domestication through separation affects the lives of the animals concerned. From the very outset, the ties that bind bees to particular ecologies are ever so gently snipped. Even the establishment of the hive involved this process, when an existing hive was opened up and cut in half. At first, the new hive would have been sealed with tape to protect it from pests and predators. And the first foragers flew out into a world dramatically different from that of their ancestors. In its way, this is a separation from the more difficult food-finding environment of the Australian bush. Once established, the hive was sealed up and transported four hundred kilometers south. I set it above the level of backyard flooding, in a place protected from the cold southerlies of winter and the blazing heat of the summer sun. In fact, the hive itself is a kind of separation, because careful thought goes into its construction, such that insulation and pest-excluding properties are maximized. The hive separates bees from the need to find a hollow tree. I set the hive on an ant cap so as to isolate it from the crawling insects of the garden. And given the chance, I would have isolated the hive from the ever-present syrphid fly, who hovered over it like a cloud. Considering this hive's history, the act of feeding its inhabitants is a logical consequence. It is a separation of the bees from their feeding ecology, from the times of flowering plants and the scarcity of winter. I think that a Yolngu person reading this might be laughing pretty hard. Or frowning pretty severely.

The next month I get a call from my friend Wendy, who lives just to the west in a small community near Boggy Creek. She knows about my native bees, and she thinks I might be interested in what's happening on Boggy Creek Road. The local council has cut down a tree that was threatening to collapse on the road, and in doing so they've exposed a nest of stingless bees. "Those poor bees," Wendy says, anticipating that the colony has met its demise. But I happen to have an empty hive box that Tim gave me to facilitate a hive split later in the summer. I load it into the back of my car with some tools and drive to a shady bend in Boggy Creek Road, where I find a felled tree that appears to have had half its trunk sheared off. The bees are instantly apparent—there is a throng of activity around the exposed nest where their futile emergency procedures have been initiated. I can also see two syrphid flies hovering above the nest, no doubt spoiled for choice as to where they can lay their eggs. In bee terms, it's a tragedy upon a disaster.

I take my empty hive box from the car and lay it on the ground beside the fallen tree. Using a metal scraper, I scoop out as much of the brood as possible and, with honey and bees all over my hands, place it as centrally as possible in the box. A local resident drives past, and when he sees what I'm up to he smiles and gives me a thumbs-up. People are concerned about native bees, and as with dingoes they see the act of domesticating them as a positive thing. I anticipate that the bees will need some propolis to rebuild their hive structures, so I scoop out as much as I can and place it beside the brood. As for the honey stores, I collect these and put them in a plastic bag that I'll give to Wendy as a finder's fee. Interestingly, the honey harvesting is the most enjoyable part of the process for me, and while I slurp the sour honey off my fingers I feel a faint connection to the bees' cousins in Arnhem Land and the Yolngu people who slurp likewise. Having salvaged as much as I can, I reassemble the hive box and insert a sheet of clear acetate that will serve as a viewing screen. That way, I can check on the bees' recovery. I load the hive into the car and take it from the forest to my suburban backyard.

Having found a suitable shady, sheltered spot for my salvaged hive, I mount it on a pedestal and leave it for a week to give the bees a chance to recover and rebuild. I can't imagine that more than a few hundred bees accompanied me with the salvaged nest, so I expect that it will take a long time for the colony to become as strong as it once was. I don't even know if I managed to salvage the queen with the brood cells. If not, then I can only hope that the depleted colony will produce a virgin queen who will go on a mating flight and return to lay eggs. After a week, I take off the top section to see what's happening within the hive. The acetate sheet is hard to see through, so I remove that as well, anxious that I not expose the hive for too long. I'm surprised and encouraged at the amount of rebuilding the bees have done; they've made a lot of progress on a new labyrinth of propolis pipelines and have even extended some structures up to the second section. I quickly snap a couple of photos and close the hive.

Looking through the photos on my computer screen, I notice in one, some white things in a corner of the hive box. I zoom in on the photo and my heart sinks, as there on my screen I see a cluster of more than fifty squirming maggots. I go to Tim Heard's bee book and identify them: hive beetle larvae. So it wasn't just syrphid flies who were threatening the colony—there were hive beetles as well! I was prepared for this outcome but had vainly hoped it wouldn't happen. In hindsight, given how long the nest was exposed on the road, I suppose it was inevitable. But having gone to all the effort of salvaging the hive and investing myself emotionally in its survival, the maggoty threat we now face is devastating. And I find myself at a domestication crossroad. I've taken a colony from the Australian bush and brought it to the suburbs, where the bees will forage on strange, wonderful, bountiful plants. I've taken a path toward unmaking, toward the separation and protection of bees that will markedly distinguish them from their former lives in the bush at Boggy Creek. But the choice I now face is not whether to intervene and remove the beetle larvae. The hive is so weak that it would surely be devastated by so many wriggly worms. The choice before me is whether or not to domesticate these bees. If I do not intervene, it will spell the end of the salvage project and the colony will die. It's like being on a treadmill with no "off" button. You must keep running with it, because if you stop, you'll be thrown off. I go to the bathroom to fetch a pair of long-handled tweezers.

4.

Crocodiles

It takes a civilized, cultured, overcrowded man to hate crocs, or love them, or exploit them, or exterminate them.
—ALISTAIR GRAHAM, *Eyelids of Morning*

What is a crocodile worth? In monetary terms, quite a lot. A top-of-the-line crocodile handbag from a well-known brand retails for $52,500. The full-grain Taurillon (calfskin) leather equivalent is only $5,600.[1] I know entire families in Ethiopia who will not see that kind of money in all of their lifetimes combined. Those handbags must have lots of storage. But does the price reflect the value of the crocodile, or is the bag just a bit of crocodile who has been reduced to his component parts, with the value added by the application of a chrome logo? The cost of the bag doesn't tell us a lot about how much a crocodile is worth in meaningful terms. It says there's money to be made from brand names, bags, and crocodile production, but can a crocodile's worth be reduced to the price of a handbag?

Ironically, it was the value of crocodile skins that lent more worth to live crocodiles than to dead ones. The demand for fashionable handbags after the Second World War put a lot of pressure on wild crocodile populations in Australia. In the north, ready access to guns, spotlights, and outboard motors facilitated such an assault on the crocodile population that some people feared they would be wiped out by the end of the century. Up until that point, crocodiles were not held in particularly high esteem—at least not by colonial Australians. All they did was kill cattle and people, and in a climate that beckoned folks into the water, they made swimming a precarious affair. Crocodiles were vermin. But by the end of the 1960s, so diminished were crocodile numbers that the governments of Western Australia and the Northern Territory passed legislation protecting saltwater crocodiles.[2] The government of Queensland was not as quick to act. Notoriously corrupt and hell-bent on resource extraction, it had no inclination to protect crocodiles.[3] But in the end it didn't matter, because in 1972 the federal government prohibited trade in crocodile skins. Queenslanders could continue to kill off their crocodiles, but the primary motivation for doing so—profit—was effectively removed.[4]

In 1979, crocodiles were moved to Appendix I of the Convention on International Trade in Endangered Species of Wild Fauna and Flora (CITES), which was effective in stopping the decline in saltwater crocodile numbers worldwide. Then, as numbers began to recover, they were moved to Appendix II, which allows for some trade. At that point, crocodile farming was seen as an ideal way to reconcile crocodile conservation with trade in crocodile skins. This is in line with the principle of sustainable use.[5] David Quammen calls this the "use-it-or-lose-it" principle;[6] it's based on the premise that if there is no practical (i.e., monetary) benefit in maintaining crocodiles, there will be no motivation to ensure their persistence in the wild.[7] So an economic incentive motivates people to farm crocodiles, and a corollary of this is that their numbers in the wild should increase. There are two means by which this is supposed to work. One is crocodile ranching, in which eggs are collected from the wild and hatched and raised on farms. This gives landholders, both traditional and colonial, a motivation to harbor crocodiles on their land, and by providing a legal outlet for skins, it discourages black-market trade, which relies on poaching. Crocodile hatchling success is also quite low in the wild, so taking the eggs and hatching them in captivity is a more reliable way to increase numbers of wild crocs, provided that some are released into the wild. The other means of conservation through sustainable use is farm-rearing, in which crocodiles are bred in captivity; their eggs are incubated and the hatchlings raised to maturity, at which point they are slaughtered. This "closed" form of production also provides easily accessible products that are quality controlled, so proponents argue that it promotes conservation by easing pressure on wild populations.[8] Some farms also host tourists, and this is supposed to raise awareness of crocodile conservation issues. What the tourists do with their newfound awareness is anybody's guess.[9]

In the case of Australian saltwater crocodiles, the shift to sustainable use marks a profound change in the way in which people engage with these animals. Conservation of species, while well intentioned, holds on to the hem of late modernity like an awkward child. The notion that a particular population can be reduced to a singularity based on genetic relatedness, and that conserving individuals of this species somehow maintains the intactness of the world, is a complete and utter denial of the complex connectivities of ecological systems. This is why zoos have been criticized for claiming to preserve species as genetic material, when in fact their inmates have as much relation to their ancestors' ecologies as Jersey cattle have to aurochs. Now zoos have to justify their incarceration of animals with claims that they are raising public awareness of conservation issues. This is apparently something that cannot be done without keeping animals in captivity.

It was the moderns who conceptualized a separate, protected "nature" and set about creating it. National parks have boundaries for a reason. They mark a profound separation in conceptual terms of humans from places that are seen as natural. Where humans are an inconvenient presence, as with Bushmen in Botswana, they are pushed out or incorporated into the systems as tourist attractions.[10] In the grand tradition of late modernity, conservation neatly shifts from concrete to abstract reference points and then back to a much narrower version of the concrete.[11] At the conservation level, crocodiles are abstracted. In one sense, it does matter whether a particular crocodile in a particular place is protected, and conservation measures might work toward that end, but conservation is essentially a numbers game. Most important is the number of crocodiles and whether that number is increasing, stable, or decreasing. This in turn largely determines life and death for particular animals. If crocodiles are abundant, then losing a few individuals is no big deal; if the number is critically low, then every individual counts. But at the conservation level, the individual is subservient to the species. This is why conservation language is replete with the definite article and a singularity. It is *the* saltwater crocodile—singular—that must be protected; it is *the* saltwater crocodile whose habitat is threatened. In conservation discourse, there is no plurality of crocodiles, just a number attached to a species, referred to as a single entity and made conceivable through our linguistic cultural frame of reference. But when *the* saltwater crocodile is moved to Appendix II of CITES and eggs are collected to be hatched and reared, then objectification and singularity become very difficult indeed. A five-meter crocodile weighing close to a ton is a difficult subject to objectify. I'll explore this later because it speaks to the worth of crocodiles in very concrete terms.

As it happens, Natasha Fijn did not limit her study of Yolngu ecologies to human relations with bees. She was also interested in how Yolngu people related to crocodiles—in no small part because her friend and colleague Val Plumwood was attacked and badly bitten by a saltwater crocodile in Arnhem Land.[12] Val's experience and her writing on the subject impelled Natasha's idea that ecologies are not simply hierarchical, with humans at the top and bacteria at the bottom. She came to understand Yolngu perspectives that see people as entangled with other creatures, from crocodiles to maggots, all of whom are important within Yolngu ecologies. Debbie Rose neatly sums up this perspective in her Indigenous philosophical ecology: "Life is for itself and also for others." So while a person might actively avoid being eaten by a crocodile, if such a thing should happen it is not an aberration or the crossing of a boundary that challenges the view that humans occupy the apex of life processes. In the Yolngu ontology, some are eaten by others, and regardless of their tools, dances, and cunning, humans are not exempt from that rule.

Back in the engine room of academia—the university café—Natasha explains to me how Yolngu people blend the singularity of crocodiles as a species with the plurality of individual crocodiles. For the Yolngu, crocodiles as a species have what wider Australia might consider an odd association: they are nurturers. Tash explains, "Wider Australians tend to think of crocodiles as something like a dinosaur: aggressive, unpredictable, and kind of evil. But for the Yolngu, she's a nurturer of her eggs and her nest. This is also symbolic of nurturing children. That's why at the primary school in Yilpara the emblem is a crocodile guarding her nest and her eggs, because that's what the school's about: nurturing and protecting children." As individuals, crocodiles represent both species and selves as well as ancestors and associations with places. Tash tells me of a Yolngu woman, a custodian of the community, for whom crocodiles were totemic. "For her, the crocodile as a totem was really important, and that made individual crocodiles significant. She'd take note of a crocodile that would cruise up and down the river and keep track of where he was." There were multiple reasons for this woman's concern for the crocodile. At one level, he was an ancestral figure for and to whom she had certain responsibilities. At another level, there was the issue of personal safety: "You have to take note of the crocodile because that's a significant figure in your life, not just in totemic terms. Even if you don't see it, you're still thinking about that crocodile quite a lot, because otherwise you might get eaten." Tash gives another example, this one from a mining township called Nhulunbuy. On weekends, miners and their families like to go to the beach to swim and fish. But there was a resident crocodile who used to cruise back and forth in the water, and he made these beachgoers a little nervous. Some local officials had the crocodile captured and relocated to a wildlife park in Darwin, but the poor creature died in transit, probably from a fatal buildup of lactic acid as he thrashed about in his restraints. What was offensive to the local Gumatj elders was not just the death of an animal with totemic significance but his removal from Gumatj clan land. The crocodile as both totem and individual was intrinsically connected to the people, not just by being a crocodile but by being in that particular place.[13] While his death was offensive, it was a greater offense to remove him, akin to tearing apart the ecological integrity of the land.[14] "For individual Yolngu," Tash explains, "it all comes back to that piece of land. If you can always link it back to place, then that's where the grounding is for you. If you belong to that land, you can work out all your structural associations from there." In the end, the authorities placated the Gumatj elders by having the crocodile preserved by a taxidermist and returned to clan land for a funeral ceremony.

Tash's mentor, Howard Morphy, describes another incident in which the life and death of a crocodile and a human were linked to a particular place. Morphy was with a Yolngu hunting party at Yalangbara, the place that marked the boundary between the territories of the two stingless bee moieties: Yirritja and Dhuwa. The leader of the Dhuwa moiety clan saw a crocodile swimming in the sea below and raised his gun to shoot. But in a moment of unscripted drama, a Yirritja woman placed herself in front of his gun, ostensibly protecting the crocodile. There followed a "heated discussion," after which the man lowered his rifle.[15] As Morphy came to understand, the crocodile had crossed the boundary from Dhuwa sea into Yirritja sea (yes, territories extend into the ocean), and the crocodile in the abstract is a major ancestor of the particular Yirritja clan associated with the land's creation. As Morphy explains, under normal circumstances it would have been permissible for the leader of the Dhuwa moiety clan to shoot the crocodile, but at the time, the leader of the Yirritja clan was ill. It was the Yirritja clan leader's daughter who placed herself in front of the rifle, and she did so because she believed that shooting the crocodile might weaken her father's spirit. Yet it was not just the association of a particular crocodile with a clan and its leader that led her to take this action. While they were discussing the issue, the crocodile turned and swam back into Dhuwa moiety territory. The man raised his rifle again, and this time the woman did not intervene. The crocodile no longer held an association with the Yirritja moiety's territory and was therefore no longer as closely associated with her ailing father. But in a moment that verged on comedy, the indecisive creature turned again and swam back into Yirritja moiety sea. Once again the hunter—not without some frustration, I'd imagine—lowered his rifle. So here we have a world in which the worth of a particular crocodile is not a factor in some global series of numbers or the fluctuation of some particular market value but instead is directly connected to an ecological network that includes humans. The woman who stood in front of the rifle saw herself as an agent in an ongoing process that she inherited from her ancestors. That process associated her lineage with a place and the creatures, including humans, who occupy it. Her concern was for the health of the land of her ancestors, and so when a threat was made to a creature associated with her ill father, the land itself was threatened, and she was compelled to act to prevent harm to a particular crocodile and a particular family member. But when the crocodile crossed into a place with which her association was not as strong, it crossed into abstraction and its worth changed. For Yolngu people, crocodiles slip from the abstract to the concrete as smoothly as they slide from muddy bank to water; the underlying ontology has nothing to do with the value of crocodiles, either in abstraction or as concrete individuals, and everything to do with their value as integral parts of places and, by extension, ecologies. It is this same perspective that allows these people to destroy nests of bees with an underlying intention of keeping the land healthy. As such, a crocodile cannot be unmade in the colonial sense; she cannot be valued in isolation from the place she occupies.

We can get a sense of how Aboriginal people of former times conceived of crocodiles from the stories they bequeathed us. It's our good fortune that two Australian anthropologists saw fit to record as many stories as they could in as accurate a manner as possible and put them all together in one collection. Ronald and Catherine Berndt of the University of Western Australia spent almost fifty years working among Aboriginal people throughout the Kimberley, Arnhem Land, and the Central and Western Deserts. They scrupulously recorded and translated hundreds of stories, ranging from the all-encompassing "How the Milky Way Was Formed" to the exquisitely titled "Disintegration of Tick Woman."[16] The Berndts were disappointed in the ways in which colonial retellings had traditionally rejiggered and censored

Aboriginal stories. The stories presented in their book are "unexpurgated," or in other words uncensored and as close as possible to the original telling, with the naughty bits laid bare.[17] They also break with Western tradition in that these stories are not credited only to language groups but also to particular named individuals within those groups. More than seventy storytellers, male and female, are credited in their collection. According to the Berndts, these people were concerned that their stories be preserved for posterity. They went to great pains to ensure that the Berndts understood the stories, so that they could be understood as the record not just of a people but of a people who were intimately connected to the land, the creatures of the land, and the cosmos. There are many stories of crocodiles in the collection, mostly from Arnhem Land and the Goulburn Islands. The majority of them are not so much about how crocodiles became nurturing parents but about how they became so dangerous to people. They begin with the crocodiles as dreaming beings. Like the dingo children of the story in chapter 2, they are essentially humanoid in form but prophetically named after the creatures into whom they will inexorably be transformed. In one story, told by a Gagadju man named Fred Wadedi, the crocodiles are the sons of two women.[18] While their mothers are collecting lily roots, the sons travel farther afield to hunt goannas and snakes. Before the sons return home, the women finish collecting food and take it back to camp, where they make a fire, cook the food, and eat it. Just before the sons return to camp, the women inexplicably take the fire and hide it in their vaginas.[19] The boys ask what happened to the fire that they saw from far away, and the mothers deny having had a fire. Implicit in this story is a social prohibition against the boys' making their own fire, so the boys are unable to cook the animals they've caught. They go hungry and the meat rots. The next time the boys go out hunting, their mothers do the same thing. They cook and consume all of their own food and hide the fire in their vaginas before the boys arrive back at camp. It has been raining, so the boys return cold and hungry, but with no way of cooking the game they've caught and no way to get warm and dry. The next morning, the boys go out again, but this time they are less interested in hunting than they are in modifying their bodies. They whack each other's jawbones with sticks until they are elongated like those of crocodiles; they take gum from an ironwood tree and fashion crocodile noses for themselves; they jump into the billabong to see how well they can blow bubbles underwater. At this point, they decide to kill their mothers. They lie just below water level in the billabong and wait for their mothers to come collecting lily roots. As their mothers wade into the water, the half men, half crocodiles attack, dragging their mothers under the water and cutting their throats. Despite having crocodile heads, they can still speak: "We will kill animal, man, anything!"[20] They've positioned themselves within the wider ecology as top predators. They take mangrove sticks and break each other's limbs, making them short like those of crocodiles. The transformation is complete.

There are two other stories in the collection that follow a similar theme. From a Maung man named Peter Namiyadjad we are told of an ancestral Maung crocodile man, Gwunbiuribiri. He's staying on the King River when a group of bird people arrive. Together, they travel to the coast at a place called Wandjili, from which two men begin to ferry the bird humans across to Goulburn Island. Gwunbiuribiri asks to be taken across with the birds, but the ferrymen tell him he'll have to wait. They tell him he's too heavy, but in truth they don't want to ferry him across because he's dangerous. Gwunbiuribiri stews over this rebuff and plans his revenge. Like the two brothers in the previous story, he extracts some gum from an ironwood tree, which he molds and presses onto his neck and nose. He calls his friends the goannas and suggests that they go down to the creek to see how long they can stay underwater. They all dive down, but one by one the goannas are forced to surface and breathe. Only Gwunbiuribiri is able to stay under the water. He tells the goannas—even Water Goanna—that they are better off on land. He swims across to Wandjili, where the two men are ferrying the last of the birds across to Goulburn Island. Seeing this, he dives under the water and surfaces just behind the canoe as it nears the shore. He takes hold of the canoe and tips it over, spilling the bird people into the ocean. As they flap about in the water, they become birds and fly into the sky. Gwunbiuribiri declares, "I'll make myself crocodile. If I meet any people, I'm going to kill them; it's better for me to be a crocodile!"[21]

In another story, this one told by a Ngulugwongga man named Mathew Maluwau'wau, the crocodile man is named Yingi. He lives with his two wives beside Bangeran billabong, and one day he sees some men go hunting for ducks. They catch a brace of ducks and return to camp, where they cook the ducks and begin dancing. Yingi joins in the dancing, fully expecting the men to share some of their duck harvest with him, but they do not share. Night falls and Yingi returns to his wives, complaining about this poor treatment. The following day, the men once again return with a brace of ducks. Again they dance, and again Yingi joins in but is not rewarded. Anger begins to boil within him. The following morning, he cuts a club from a bloodwood tree, paints it white, and gives it a red ochre tip. He watches the men return from hunting and once again joins in the dancing. When he stops dancing, the men cry out for more but instead he takes his club from its hiding place. The men run from him in fear, but he catches and kills at least five of them. On returning to camp and telling his wives, they berate him, asking, "Why didn't you kill them all?" He replies that he did, but we suspect that he's lying and that some of the men have escaped to warn others. The following day, Yingi's wives encounter another group of men at a place called Djinan-derara. The men give them catfish and barramundi to take to Yingi, along with an invitation to come and stay with them. Yingi goes the next morning and decides that this is where he will make camp. The men point to a fish trap—a woven basket that allows fish to enter but prevents them from getting out. They tell Yingi, "That belongs to you, and it is full of fish!" Yingi can't believe his good fortune. He pulls out as many fish as he can and then climbs inside to get the rest. The trap is sprung. With Yingi hooked on catfish barbs inside the trap, the men roll it into the Daly River. One man on the other side of the river transforms himself into a dog, dives into the river, and swims over to the submerged trap, presumably to rescue Yingi. But Yingi in his rage drowns the dog. He surfaces in the form of a crocodile with the dog in his jaws and rails at the humans watching from the bank, "Why did you throw me into the water!" He forgoes the path of the dead—who become "dreamings," physical features of the land—and instead chooses the way of the crocodile. "Nobody come swimming in this river! Look out for me! You people, dogs, wallabies, and other creatures, be careful! You will all be frightened of me!"[22]

Undoubtedly, these stories speak to the histories of particular groups of people in relation to particular places. As in all Aboriginal stories, the land is the "common denominator," that by which histories are grounded and contemporary relations come to make sense.[23] As the Berndts make plain, the land "speaks" through the dramas that are enacted in particular places. This is why storytellers pepper their tales with references to very specific places. In each of the three crocodile stories, particular places hold the origins of relations between crocs and humans. And in each case crocodiles become dangerous to humans through the acts of their ancestors. Crocodile is slighted, refused food, disregarded, feared. The humans in the stories are afraid of the dangerous ancestral crocodile/man, but they invariably mishandle the situation. They delay, they set the problem aside, they abide him only to a limited degree in the hope that he'll go away. And in each case they get it utterly wrong. Their feeble, passive snubbings only make him angry, vengeful, murderous. Through their relative inaction, they set in motion the train of events that will lead to an enduring danger to humans and other animals. These stories place responsibility for the ways of the world squarely at the feet of ancestral humans and at the same time engrave into particular places the rights of crocodiles, not just to exist but to kill people. Thus at the end of each story the crocodiles make it clear that, given the opportunity, they are going to kill. You have been warned. But the stories also speak of another characteristic of crocodiles: they and humans come from the same stuff. Their dangerous nature made them an awkward fit for society, but the ancestors made it so. These crocodile men fulfill their destinies and eventually metamorphose, but even after their transformation they speak and make threats. Indeed, in other stories they threaten to cross the species boundary and overturn the relationship.

From Guugu Yimithirr man Tulo Gordon, we have the story of a big old crocodile named Ganhaarr who lived close to an Aboriginal people's camp on the Endeavour River in Queensland. The women of the camp go daily to the river to swim and hunt for mussels without any trouble from Ganhaarr. But one day a woman takes her baby with her and sits on the bank at the waterline. Old Ganhaarr swims past all the women who are bathing and heads straight toward the woman with her new baby. He rises out of the water, grabs the mother and child, and throws them onto his back, then dives back down, taking his captives to his hollow under the riverbank. There he makes the woman his wife. But Ganhaarr proves to be a jealous and suspicious husband. He takes his wife and adopted child to sit on the muddy bank while he sunbathes, but he always keeps one eye on his wife. If she tries to sneak off, he catches her and brings her back to his riverbank hollow. One day the sun is particularly hot, and Ganhaarr falls asleep, with both eyes closed. The woman grabs her baby and runs. Ganhaarr wakes and chases her, but this time she makes it back to her people. The men at the camp take up their spears and go looking for the old croc, but he eludes them. What they do find is a trail of crocodile eggs laid by the woman during her escape.[24]

Catherine Berndt collected a comparable story from Milingimbi Island, off the coast of northeastern Arnhem Land. In this story, the crocodile ancestor similarly steals his human wife, but rather than being jealous and possessive, he tries to be an affectionate and dutiful husband and father. Inevitably, this mixed marriage is doomed to fail; the woman escapes and the crocodile weeps over the baby crocodiles born to the woman and left behind.[25] And Jeffrey Heath relates a Nunggubuyu story from southeastern Arnhem Land in which a crocodile steals a woman from her people and takes her to his home.[26] He works his tail off catching fish for his wife, but she too escapes and returns to her people, who find and spear the croc. As with the other captives, this woman gives birth to baby crocodiles.

The messages in these stories are about the rules of marriage and the importance of staying with your own people. Although the women are kidnapped, they all strive to return to their own kind. But these stories are also about crocodiles. They describe an underlying, enduring personhood in crocodiles based on shared ancestry.[27] Like other animals and features of the landscape, crocodiles share a common descent with humans, and their ancestry endures; they might be cunning or vengeful, they might desire human brides, they might prove jealous and possessive or devoted and caring. In any case, they have needs, wants, motives, intentions, feelings, memories, and histories that frame their relations with humans. Contrast this with a typical biology textbook: The crocodile is small-brained and cold-blooded. It lies in wait in a state of torpor until a clever mammal happens along and the vibrations of this warm-blooded prey trigger the predatory instincts of the ancient and implacable monster. It launches from the water like a sprung trap and the mechanics of its jaw musculature close on its victim, which it drags into the water and spin-cycles until the mammal drowns. Little sensors on the sides of its mouth compel it to snap its jaws sideways and loosen up the meat with its teeth so that it can swallow, whereupon the mechanics of digestion turn the food into energy enough to catch its next meal. Sated, it returns to its state of torpor. Anyone familiar with crocodiles will readily sign off on this account of crocodile predation. But anyone who farms them will want double spacing, because there's a lot of subjectivity that needs to be written between the lines.

I love causeways. Not so much the ones that span rivers but the ones that bridge islands with the mainland. I love ephemeral causeways, like the old one at Le Mont Saint-Michel, where the tides give you a temporary pass to cross and then swamp the causeway and hold you on the island until the moon lets you go. It's like a slow-motion game of chicken with the ocean and its master. Every day, causeways transform islands from places you can visit to places of refuge. They give islands a sort of accessible mystique. This is amplified when there happen to be five thousand crocodiles on the island. Mudflats surround the bump of land that harbors Bindara crocodile farm and the three-hundred-meter causeway that affords the only way in. The farm sits amid mangroves and mud on the coast of tropical Queensland. The neighbors probably appreciate that there is only one, tide-dependent escape route from the farm, though a saltwater crocodile would see things differently.

Pulling into the car park, the most obvious feature of the farm is the crocodiles. They bask in the sun behind chain-link fencing. Others are semi-submerged in their ponds, and I presume there are some who are completely concealed by the water. With crocodiles, you never know. Having assured myself that none of these massive reptiles is on the same side of the fencing as I am, I take my eyes off them and survey the rest of the farm. There's the usual accumulation of farm machinery, old cars, earth-moving equipment, and piles of scrap that might come in handy one day. The parking area is full of vehicles, and it's easy to tell which belong to staff and which to visitors. The staff cars have months' worth of dust layered over the paint; the visitors' have a light dusting from the road in, or from the well-traveled, sunset-colored dust of inland Australia. The out-of-state license plates are another giveaway. Beside a couple of RVs, there's a tour bus. The passengers are climbing aboard after their farm tour. Meanwhile, other tourists are entering the visitor center. I'm here to do some in-depth study, but for the sake of an introduction to the place, I'm going to tag along with the farm tour. I've even brought my wife, Tigi, and our two-and-a-half-year-old girl, Leni, along. Leni is predictably fascinated by the "tot-awhiles" and firmly attached to my leg, from behind which she peers at the huge, toothy animals.

The building we enter is an expansive, timber-framed hall with a host of adjoining rooms. The main lecture hall is filled with tables and chairs; it doubles as a restaurant. The walls and ceiling are decorated with all manner of crocodilia: cabinets hold crocodile skulls, teeth, handbags, and belts; taxidermied crocodiles are mounted high on the rafters, and crocodile skins adorn the ceiling. There's also an intriguing poster-size photo of two New Guinean men in a canoe, their chests scarified in a way that is remarkably evocative of crocodile scales. We go over to reception to sign up for the tour. Here are the souvenirs: T-shirts, beer coolers, pens, wallets, and cuddly plush crocodiles that remind us exactly of what crocodiles are not. I introduce myself to Karen, the tour guide. She's had a heads-up that I'm coming, so she spends a bit of time giving me some background info. Karen is small of stature, with hair tied back and a tattoo on the inside of her forearm. She's just completed her PhD in geology, but jobs in the mining industry are drying up. Such is the job market in rural Queensland that Dr. Karen finds herself working on a crocodile farm.

Karen and I return to the main hall and she begins the tour. Twenty visitors are arranged around the tables, and the diminutive tour guide cranks up the volume and her Aussie twang. She begins with a conservation message, telling us what the status of wild crocodiles was when the farm was established in the 1970s and their status at present. Since then, fourteen crocodilian species have been taken off the critically endangered list, and not because they've become extinct. Crocodile numbers are recovering worldwide. Karen moves from crocodile numbers to the subject of crocodile meat. She describes the different cuts and the particular cut you can expect to eat if you dine at the restaurant after the tour: the tail. But, as she says, the meat is a by-product. It is the skin of a crocodile that makes farming worthwhile, at least in monetary terms. She holds up a piece of crocodile skin, dark brown, scaly, knobby. "Now, guys, a piece of skin straight off the crocodile's back is worth twenty-five dollars U.S. per linear centimeter. This is what we call the back strap." She indicates the bumps. "These bumps are called osteoderms, or scoots. They are pure bone and they're fused to the crocodile's spinal column. We've seen bullets bounce off the back of a three-meter croc. You'll notice if the crocodiles get into a fight, they'll turn their backs to their opponents, to where most of the protection is." To demonstrate just how tough these osteoderms are, Karen picks up a soup spoon and whacks it against the skin until the spoon is bent and the skin undamaged. I make a mental note never to attack a crocodile with a spoon. Meanwhile, Karen holds up a piece of belly skin the size of a hand towel. It is pale and smooth and not the sort of skin you would expect to see on a massive, knobby, greenish-brown monster. "Across the chest—a piece like this—three thousand dollars. This is the most valuable part of the crocodile." She goes on to explain the value-adding process from a handbag sold at the farm, to a no-name bag sold in a chain store, to one with a brand name added and sold for tens of thousands of dollars.

As we follow Karen outside for the tour, she prepares us with some details about crocodile behavior. "You need to know what you have to run from today. But bear in mind you only have to run faster than the slowest person." She introduces us to Anthony, another tour guide. "Please do not be offended if Anthony taps you on the shoulder. It means you're protruding too far over the fence. After ten taps, we'll throw you over the fence." The tour is peppered with reminders that these are dangerous animals. As we stop on a causeway between ponds, Karen asks us to look the pond over and guess how many crocodiles it contains. "Guys, how many crocodiles have you counted in here so far?" Someone answers four. "Four? You have a big problem. You might wanna start running. We have fifteen in here: fourteen girls, and one very happy, busy boy; his name is Casanova." A few chuckles. "You can guess what his job is. And that's him right at the surface at the moment. See if we can get him to come up." Karen takes a chicken carcass from a bucket she's carrying. "On a cloudy, overcast day like this, it's a little cool for them. So some of them may not be too hungry. But these guys know through routine and repetition, same time every day, we're going to offer food. If they come up, we know they're hungry. If they don't, we know they're not interested."[28] She lobs the carcass to the big, half-submerged male. It lands beside his mouth and he nonchalantly snaps it up. "Okay, so he's hungry. Crocodiles of this size can go up to twelve months between meals. So remember. If you're planning on climbing up a tree to escape a crocodile, they know what goes up has to come down. Can you sit in a tree for twelve months without food?"

She leads us across to the other side of the path and points to a pond that looks empty. "This is Blondie. He's Casanova's brother. Please, nothing through or over the fence, 'cause he's not a gentleman and they're not gentlewomen. Oh, we're at high tide. Blondie, where are you? Come out and play with us. Blondie!" There is no sign of Blondie. The head of a smaller female croc breaks the surface momentarily and then recedes below the water. "Now, guys, a crocodile has the ability to slow their heartbeat down, two to three beats per minute. So for six hours they can stay underwater before coming up for air."

We follow the path to another enclosure where a placid pond takes up most of the space. Karen stands in front of the water and faces her audience. "All right, guys. There's one crocodile right in front of us right at the moment. It's Fat Bill. Can anyone see him? Guys, please remember, if you cannot find a croc on a croc farm, what hope do you have in the rivers and creeks?" Everyone is craning to see the invisible croc beneath the murky water. "You don't see one?" Karen drops a chicken carcass at the edge of the water and an enormous, knurled head emerges from the water like a leviathan with lumps. Fat Bill surges forward and snatches the food. "He has to bring his head up to swallow, so you'll get to see the true size of him very shortly. There's his tail. And his head is still at this log. There he goes." The crocodile before us is enormous. "Now you can see why we call him Fat Bill, can't you?" There are gasps among the people in the crowd. "Can you see him now? We'll give him one more." Karen throws another carcass, which misses its mark. "Oh, I missed him. He may get cranky with me." Fat Bill drops his chin to the water's edge and resumes his state of torpor as though waiting for someone to enter the pen and move the food closer to his mouth. "We started off with quite a few more crocodiles in here, but now he has just two girlfriends. It turns out, if a boy doesn't like a female, he kills 'em. So matchmaking is quite difficult. So we found two that he likes and haven't messed with it since." The females being discussed remain hidden from view.

Karen leads us across the car park to yet another enclosure, where a large male lies halfway out of the water. This is Nathan, a wild-born crocodile who had a habit of lying in wait under fishing boats and taking barramundi from the fishermen's lines. Karen throws a succession of chicken carcasses over the fence, but he shows interest only in the first. She bangs on the fence, trying in vain to get him to come closer. As she calls and bangs on the fence, Nathan crosses over a female crocodile lying next to him. "He's just gonna swim over the top of his mate. He's just and she's just . . . um." Karen begins to look decidedly awkward. "I'll keep it PG. This is what mating looks like. Um. Interesting that it started to happen. Crocodiles are seasonal. Mating usually happens end of November, December, January. Well, it's happening early, apparently. Guaranteed, every time I have small kids on a tour, either this happens or a bird gets taken. Let's move on." I'm reminded of an early litter of dingo pups.

We arrive at the last of the enclosures and meet Buka, the largest crocodile on the farm. We learn that Buka is 5.3 meters long and weighs more than a thousand kilograms (2,200 pounds, or more than twice the weight of a horse). Karen throws a chicken carcass and Buka propels himself up onto the bank to snatch the food. Karen throws another and he pushes himself farther. He seems to have to make an extraordinary effort to move on land. Rather than eat the second chicken, Buka simply lies next to it. His female companion moves toward the food, but Buka makes a sideways movement to fend her off. She leaves the food for her mate.

Having seen the largest crocodile, we are led to the garden behind the main hall to see a small one. We wait a few minutes and Karen returns with a baby croc, his snout securely taped shut. It's an opportunity for the tourists to have their photo taken while holding a crocodile—though not without some instructions. "Please, guys, if he does wriggle, don't drop him on the ground. It took over two hours last time to catch him. The other thing, guys:

please do not bring your hand over the top of his head. He thinks you're a predator. It puts him in a bad mood and he won't let anyone hold him." The kids line up and each is allowed to hold the taped-up crocodile while their parents take photos. The adults follow. Karen asks Leni if she wants to hold the crocodile. She refuses, which is not surprising; she's going through a stage where she's afraid even of worms. The photography session over, we are invited to visit the restaurant and eat some crocodile. I try the tempura-battered crocodile fillets, which are tasty enough, though they pose little threat to the fishing industry.

The next day, I return to the farm to visit the owner, Mike. Karen leads me through the back of the main hall to the house. The outside is unremarkable, and the cluttered office we walk through is a marked contrast to the expansive, timber-framed living room within. The walls are adorned with crocodilia and ornaments that speak of a life well traveled. Square-jawed and silver-haired, Mike radiates the sort of robustness that one would need for a life spent wrangling crocodiles. But he's also quite thoughtful, erudite, and gentlemanly. Mike grew up in the suburbs of Melbourne, attended an agricultural college, and bought his first farm, a dairy operation, at age nineteen. After some teaching and research work in Melbourne, he took a job as a wildlife research manager in Papua New Guinea. This is how he became involved in sustainable-use projects involving not just crocodiles but also megapode birds, butterflies, rusa deer, and cassowaries. The butterflies were farmed for their decorative wings, the megapodes were managed for wild harvest of their eggs, the deer were to be wild-harvested for venison, and the cassowaries were farmed for feathers used in wedding ceremonies.[29] In the end, the crocodiles proved most successful. Mike established a three-tiered system of crocodile farming consisting of village holding farms, small business farms, and large, town-based enterprises. He also developed methods for slaughter and the reduction of waste. Still, crocodile farming was not possible across the entire country. Among one language group, crocodiles were considered so spiritually powerful that it would have been shameful to fence them in. When this became apparent, Mike scrapped plans to establish a farm in that group's territory. It was in Papua New Guinea that Mike's perspective on crocodiles was altered. He had entered into the job with a notion that crocodiles were merciless, cold-blooded killers, but after he spent time learning from the locals, he came to have a profound love and respect for these animals.

After seven years abroad Mike returned to Australia, determined to establish a crocodile farm. He was committed to the sustainable-use model of crocodile management and saw in Australia an opportunity to work toward crocodile conservation while applying the skills and knowledge he'd gained in Papua New Guinea. He looked at a map and chose a location off the Queensland coast that was not as hot and humid as Papua New Guinea. Bindara lies at the southern end of saltwater crocodiles' range; the cool winters cause the crocs to slow down their metabolisms outside breeding season, so they require less food.[30] Ideally, the breeders would be maintained at Bindara and their progeny "grown out" on a satellite farm in the north, where the constant warm climate would ensure a faster growth rate. When Mike set about applying for permission, he found that there was no legislation in place for crocodile farming in Queensland. He spent eighteen months seeking permission, local council approval, and public acceptance before he was eventually allowed to farm freshwater crocodiles. This was not ideal because the skins are not as valuable as those of salties, but it proved a stepping stone. After being granted protected status in Queensland, the saltwater crocodile population was booming, and crocodiles were beginning to encroach on the human population. Mike was asked to capture and remove problem saltwater crocs, and these became the founding breeders at Bindara.

Mike explains to me just how long it took to get from those first breeders to producing marketable skins. "Without egg collection in the wild, we had to start off with breeders and . . . you don't get any returns for about five years. Let's say we catch a couple of crocs now, a male and a girl, right? You can't throw them straight in together, because one will kill the other. You get them used to fences and you get them used to captivity. Then you put them together for a year, but with a fence dividing them so they can talk and communicate. Then you open a gate and let them in together. All of a sudden, you're getting some reaction. We're already a year down the track. Now, the next season . . . they'll probably mate and lay fertile eggs. Then we're three years away before we get skin production. Four to five years of no returns from your investment in breeders." For Mike, opening the farm to tourists was more a matter of survival than of simply adding income. Besides, Mike's background in teaching gave him a passion for passing on his knowledge. He loves upending people's preconceptions about crocodiles.

According to Mike, wild-caught crocs adapt well to their new enclosed surroundings because they're naturally territorial and, as I found out on the farm tour, they can tolerate quite small territories. "The first behavior we notice after they're brought here," he tells me, "is they go underwater and . . . you don't see them for a week. You might see a nose come up and take a breath—not their eyes, just the nose, and just the button—and then go down again. Then, a week later, you'll go out in the morning to do the rounds, have a look at the crocs in the pens, and you find these tracks all the way around the fence line. They got over the trauma of being caught and being transported and all the things that stress them out. They're now starting to try and determine their territory. They've got these two glands underneath the chin and beside the vent, beside the cloaca. They produce a pheromone. They lay this pheromone on the ground. We can't smell this, but they can, always. Once they've marked their territory, they're disinclined to go beyond it. Now, they define their territory quite well. We've had occasions here where our staff have left a gate open, and the crocodile walks right past an open gate at night and doesn't come out, simply because that's his territory." It's almost as if farmer and crocodile come to an agreement about the size of their new territory. And the crocs do get attached to their enclosures. "I had a crocodile get out of here, one called King Wally. You probably met him yesterday. He found a little hole in the fence and he kept pushing and pushing and he made a hole this big and he got out. I got up at daybreak and, just as I often do, had a wander down the road and a look around, and there he is on the road. *Oh, bugger*. Anyway, I roped him and tied him off to a fence post, and I thought, 'Oh, I'll go home, get the excavator, then I'll be able to manage him.' You can't handle crocs that size on your own without machinery. Anyway, by the time I got back, he'd broken the rope. Now, along the whole fence line over there, there was a little hole in the bottom corner of his pen, and he found that hole. Walked straight back to that hole and went back into his pen."

In the rivers and estuaries of northern Australia, the territories of male crocs overlap. A nine-hundred-kilogram male might extend his territory along sixteen kilometers of river, but that doesn't mean that other males are excluded. It just means that the big guy has right of way. If the big male comes cruising up the river, the other males will avoid him as best they can, but if one is a little slow to do so, then he can expect to be attacked. But this territorial network of multiple males means that a big male croc does not have exclusive access to females on his beat. While he's courting a female up at mile ten, a subordinate male will be mating with a female at the other end of the territory. As a result, clutches of eggs in the wild have mixed paternity. When Mike established the farm, he set up an expansive system of ponds with multiple males and females per the wild crocodile paradigm. But it didn't take long for the big males to rearrange things. "We've got lakes out here and we used to have sixty females and five big males. Then we had four big males. Then we had three big males. We separated the last two males and divided the lake up into smaller units." The males weren't escaping or dying off; they were being killed by the dominant male. They just didn't have space enough to escape. "Now we have breeding units where we might have ten females and one male. We keep the males separated, because they spend more time fighting than loving." There is no question of letting the males sort out a dominance hierarchy for themselves, because in the confines of a single lake they will naturally tend toward a hierarchy of one. For a farmer, this is unacceptable. Crocodiles are slow-growing animals with a lot of value attached to individual breeders, a markedly different situation from something like beekeeping, where you can afford to lose a queen because she'll quickly be replaced.

Separating males might solve the problem of male fighting, but a crocodile farmer also needs to finesse male-female relations. Some male crocs are incurable, murderous tyrants, and it takes a particular female to make for a harmonious relationship. In fact, a harmonious relationship is in many cases a pipe dream, and the best Mike can hope for is that the female survives. Sometimes it isn't even enough to give the crocs a long honeymoon period to get used to each other. "We got a crocodile here called Gucci. He's got a fairly large pond with his girlfriend in it. He's a bugger of a crocodile. He's very aggressive. . . . If she's too close, he'll smash her over the head. She's always nervous and always twitchy and always getting out of his way. But they do mate, because she lays fertile eggs. It's good clutch sizes with her. We're very happy with the result but not happy with the behavior. She is under constant stress all year. If they were in a smaller pen, she'd probably be dead. It's just that she's got the opportunity to get away from him." And then there are some males who simply cannot be reconciled with females. "We've had other crocodiles, like Boss Hog and Rasputin, absolute horrors. You put a female with him, even in a big space like one of the lakes out here, and the first thing he wants to do is to grab her and shake her and throw her around and cause a lot of damage. The amount of stress these crocodiles get is incredible. The female would run over the top of you to get away from a big male. You are not the problem; the big male is. If you want to buy a crocodile, I can sell you plenty that are males that will not tolerate any other crocodile in the pen with them, whether it be male or female."

In a farming context, there needs to be a balance between peace and productivity. In breeding crocodiles, Mike selects for certain heritable traits because they lead to increased production and better-quality skins. Some males have extra rows of belly scales. This is a trait that's passed down the male line, and it's highly desirable among handbag and shoe manufacturers. So even if a male is particularly aggressive, an extra row of belly scales is his ticket to keep on breeding. Another trait is fecundity. Among females, clutch size is heritable. Mike has found that clutch sizes tend to decrease over time among farmed crocodiles, so he selects females for clutch size hoping to counteract the decrease. "I've got seventy young females that all come from three females on the place. My average clutch size here is 41.2 eggs. These three girls lay between sixty-five and seventy every year. I've kept their daughters as my future breeders, and they're just reaching puberty now. That's gonna be interesting." Mike also selects for male fecundity. "Having a one-on-one relationship, male and female, we expect 100 percent fertility, because he's gonna mate with her several times in the two to three weeks that she's in season. We expect a high success rate in one-on-one matings. If we find we're only getting 50 or 60 percent fertility rate, that's a mark against the male." But within the biological traits associated with successful mating, there is always the need for sociable crocs: "We try and select our males for their passive nature. That doesn't always work, because I think implicit in the crocodile's nature is to be a bully and to be the boss, the alpha male. Any crocodile that doesn't kill a girl gets a tick. A crocodile that kills a female when we put them together, he's got one more chance after that. It may have been an accident. She may have been aggressive. I don't know. He gets one more. My ideal male is one that's got three extra rows of scales on his belly skin, he has a passive nature, he's highly fertile, and, generally speaking, he'll accept other females."

As for the progeny of these sometimes dangerous matings, Mike aims for fast growth rate and quality skin. The first step is removing the eggs from the breeders' enclosure. This is done not without difficulty because females aggressively protect their nests and even enlist males to help. The collected eggs are incubated in a hatching shed where the temperature and humidity are controlled to achieve maximum hatchling survival and the temperature can be tweaked to allow staff to dictate whether the hatchlings are male or female.[31] This is a useful tool for choosing breeders or males with extra belly scales. The eggs are also dipped in thiabendazole, a fungicide that is also added to the water to prevent fungal infections that crocodiles tend to get in colder climates.[32] The eggs hatch just prior to the onset of winter, so the hatchlings can't be released in an outside pond; they would die from exposure. Instead, they're kept in two-by-one-meter tubs in a heated shed. This is when the process of grading starts. "We grade about every month or two. The biggest 10 percent we take out and the smallest 10 percent we take out. We leave 80 percent that are about even. Guess what? A month later, you have 10 percent big ones and 10 percent small ones. You have to grade again. You get this juggling for position all the time." This can be seen as territorial behavior or as bullying, Mike says, "but whatever it is, it's out there. It's alive and well. If you don't grade your crocs, you're not a good farmer." Come October, the hatchlings are moved to rearing ponds, at about one hundred crocs to a pond. After growing for a few months in the first rearing ponds, they are graded and put into a second set of rearing ponds. The grading process decreases the number of crocodiles, but this is necessary for clean, marketable skins. Once the animals reach 1.2 meters in length, they go into single pens where they will spend their days until slaughter.[33]

It's mid-October, six months after hatching time, so I have a chance to see the baby crocodiles in the first set of ponds. My guide is Layne, a tall, lanky eighteen-year-old wearing a loose-fitting motocross jersey and floppy hat. He tempers his good nature with the quiet aloofness of a teen who's cultivating a distance from parents and other elders. He's only been working at Bindara for a few months. When he started, he was given the responsibility of feeding the latest cohort of baby crocs, but they were already used to another person feeding them and refused to eat. It was two weeks before they came out of the water and fed in Layne's presence. Now he refers to them as his "crockies." On our way to the ponds he breaks his silence. "There's one pond where they've got used to me a lot quicker than other ones." I ask if they're the older ones, the bigger ones. "No, they're probably the smallest ones," he replies. "They're real aggressive. They bite each other." Are they males or females or a mixture? "It's a mixture, probably. Probably mostly male, but—because you can determine the sex of the babies by the temperature you incubate them at, so they're probably 90 percent male." We pass through a gate and go around behind the breeders to a neat row of fenced ponds, each about two by six meters, separated by chain-link fencing. The ponds are lined with heavy-duty plastic sheeting and PVC pipes circulate the water. I notice two young crocs in the nearest pond slip quickly into the water as we approach. "They usually come up to the fence when I arrive with the food," Layne comments, "but they'll be afraid because of you." This is interesting. The crockies have attached themselves to a particular person. Layne understands this and encourages it. "Every day I feed the crocs at the exact same time and from the exact same spot." Do you wear the same clothes each time? "No, that changes day to day, but I wear the same hat and sunglasses. Though I think they recognize my footsteps. That's the key. I love it how they recognize who you are."

Layne will be involved with these crocs until the time comes for them to be graded and moved. At that point he'll be involved in the move to make the transition easier for the crocs. Layne leans forward against the fence and calls "Come on!" while dropping handfuls of meat at the edge of the pond. It's the same call that Karen used to attract crocs during the tour. One young crocodile emerges from the water and climbs onto the plastic sheet to feed, while the rest stay in the water, pairs of eyes bobbing about like floating knucklebones. Layne repeats the process at each pond, leaning over the fence, calling out, and dropping meat. At one pond, three baby crocs clamber over one another to get to the meat. I ask if this is the pond where the baby crocs got used to him more quickly than the others. "Yep, this is the one." I'm surprised that even I can notice a difference in this pond. I'm reminded of Marianne Lien's account of working on a Norwegian salmon farm where she had a special relationship with a particular cohort of salmon. She wrote of the salmon in cage number six that they were much more visible and responsive than the salmon in other cages. But her perception of this difference led her to respond in a particular way to the salmon in that cage. It's as if a culture of recognition creates a feedback loop in which the relationship between feeder and fed is amplified by how they perceive each other. After feeding the crocs, Layne returns to the main shed, where he's enlisted to carry boxes of frozen fish from the freezer. Compared to the intimacy of feeding his crockies, it's a mundane task.

After we return to the shed, I'm passed on like a party parcel to a man named Luke, the farmhand responsible for feeding the breeders. In ragged Akubra hat and plaid shirt, Luke looks the part. Originally from Brisbane, he left the city and worked his way up and down the state, working variously as cattle wrangler, pro fisherman, and, for the past seven years, croc handler. He's always been involved with animals, and he has a host of pets, including dogs, chickens, and snakes. We load three crates of chicken carcasses and a bamboo pole into the back of a stripped-down Land Rover. "When you're feeding crocs, your pole's your best friend. If you miss with the food, you can move it closer to the animal. Or if he gets too aggressive, you can tap him on the nasal dish. Not that they're gonna be aggressive. With you here, don't be surprised if they don't come up at all." We climb aboard the Land Rover and trundle on out to the breeding pens. On the way there, Luke gives me a few dos and don'ts. "You can take photos, but don't put your camera—or anything else for that matter—over the fence. Don't stand too close to the fence, either. There's one part of the rounds where we'll be walking between pens and it's really narrow. Keep your arms close by your sides." We stop at the first pen and Luke puts on a pair of gloves, drags a chicken carcass from the back of the Land Rover, and bends it until it snaps. "I break the keel bone on these to make them easier to swallow. Also, I add some vitamin E tablets to the females' food. It increases fertility." He pokes four tablets into one of the chickens and, holding one in each hand, walks over to the pen. Standing in front of the pen, Luke tells me, "This is Snappy Tom. Normally, he'd be up at the fence, but because you're here he's hiding in the grass." Luke calls "Come on!" and Snappy Tom emerges. He's a healthy-looking two-and-a-half-meter croc. When Luke dangles a chicken carcass over the fence, Snappy Tom lunges and Luke lets go of the carcass, which lands in the crocodile's mouth. The croc throws his head back and swallows the chicken. Having fed the male, Luke throws the other carcass to Snappy Tom's mate, who is lying in the grass by the water. I mention that everybody on the farm seems to use the same call when feeding the crocs. "Yeah, it's a call we use. Crocodiles respond really well to routine and repetition. That's why we feed them at the same time every day with the same person, same call. Say the boss came out here to feed them on the weekend. They'd look at him and say, 'Who are you?' They just wouldn't accept it." Some crocodiles also bear inexplicable grudges toward particular people. Luke tells me about a croc called Matthew who has an intense dislike for Robbie, the farm manager. "I can walk in with Matthew, walk up and back along his pen, and no worries. As soon as Robbie starts walking up that way, he'll chase Robbie up the road. As hard as he can run, and as hard as Robbie can run to get away from him. Me, I have to get up like this, stomp my feet, jump around for him to just come up and feed. But he just doesn't like Robbie." Not only do the crocodiles have to accept the other crocs they live with; they have to accept their human caretakers as well.

We approach the next pen, occupied by a croc named Ziggy, with two more chickens. Luke holds up a chicken and calls "Come on!" and Ziggy obligingly launches himself upward to take the food. When Luke throws the other chicken carcass to Ziggy's mate, I notice that Ziggy's paying attention. Luke tells me that crocodiles can feel movements through the ground and water. Whether a male tolerates a female sometimes depends on the size of the enclosure, he says. "We have a croc here called Oscar. He used to beat up his females, but he was in a small pen. Now we put him in a bigger pen he's been great. But when you see Buck—he's in one of our biggest breeding pens and he will chase that female from one end to the other. The size of that enclosure doesn't matter. He could have the whole property and I reckon he'd chase everything out of his way. She'll really gracefully slide in the water, trying to be as gentle as possible, and he'll turn around and just chase her out. I've had a situation where I've gone up there and I've startled her. She's been in the bushes, and she jumps off and splashes into the water, and as soon as she hits, he's after her like you're ringing a lunch bell. He just charges at her as hard as he can."

We climb back into the Land Rover and drive a few meters to the next pen, where we find a massive crocodile named for a Ninja Turtles character, Rocksteady. I notice that the chain-link fencing between Rocksteady and me is looking particularly mangled. I also notice him staring at me with a disconcerting reptilian stare. I've been face to face with hyenas, and their gaze has a conditional mammalian quality. There is a question in their stare; they ask, "Are you a threat or prey or something else?" But Rocksteady's stare holds no question. I ask Luke why he's looking at me so. "He thinks you're a bloody big chicken." Luke holds a smaller chicken over the fence for Rocksteady to snatch. His timing is excellent; he drops the chicken just as the huge crocodile rises upward with his mouth agape. Rocksteady catches the carcass like a pro and devours the entire thing like finger food. I mention the mangled fence and Luke takes me across the path to show me another pen, vacant and overgrown with grass. Luke explains that a wallaby once got into this pen and was bounding around trying to get out. At one point the wallaby must have stopped next to the fence, because the crocodile in the adjacent pen lunged for the unsuspecting macropod. Of course, all the croc got was a mouthful of fence, but he pulled it so hard with his teeth that the fence was distorted. Luke shows me the fence, which has a deep, cone-shaped impression where the croc pulled at it. To the uninitiated, it looks like a cannonball was shot into the steel mesh.

As we move through the farm, Luke introduces me to each crocodile. He knows them all by name. There's Snort, who runs to the fence and snorts when Luke arrives with his dinner. There's Tripod, who's missing a leg, and

Shah, who's missing half his bottom jaw. Mr. and Mrs. Woof are a happy couple, and wild-caught Toothy is Luke's favorite. In most cases, he knows these crocodiles' histories. He tells me about Yazi, who was procured from a place in Queensland where he was fed a diet exclusively of fish. On the way back to the farm, Yazi pooped in the van, which made for a stinky drive south. Luke comments on the difference between wild-born and farm-raised crocs. "The farm-raised crocs aren't intimidated by people. You can bluff a wild-caught croc with sudden movements, but not a farm-raised one. They don't have fear—at least not of their handlers." Does this make it harder to collect their eggs? I ask. "They defend them more actively—more aggressively. But we still get the eggs."

We arrive at the lake where Casanova lives with his fifteen female consorts. Luke explains how the females have their own territories, which overlap with Casanova's. In fact, it's not easy to get food to all of them because some have to cross a dominant female's territory to get to the food. He throws a chicken carcass in and the female nearest the fence darts over to collect it. He throws another onto a sand bar in the middle of the lake and a different female makes for it, but the first female rushes at her and drives her away. The subordinate croc raises her snout, a sign of submission, but the dominant female wants her gone. Two more females show interest in the food. Luke throws a carcass farther from the dominant female and one of the others collects it. The dominant female stands alert, feeling the ground, tracking the movements of the other girls. She looks decidedly put out that other crocs are being fed, and it's fascinating to see how Luke follows these relationships. Like a charity organization, he tries to distribute food equitably within a dominance hierarchy.

Back in the main shed again, I'm passed on to Robbie, the farm manager. Like anyone with thinning hair and a closely trimmed beard, Robbie cuts a strikingly handsome figure. A biologist by training, he was employed at Bindara to work on breeding better crocs for the purpose of harvesting better skins. I ask how aggression among crocs plays into selection for breeding. "There is some thought that the more aggressive crocs make for better farm genetics," he tells me, "things like growth rate, dominance, health, survivorship. But this can be detrimental to the other crocs, as you can imagine." Robbie explains that there is a limit to the amount of aggression a croc is allowed to exhibit before he's taken out of the breeding equation. "We just sold a croc to a zoo. Before we got him, he had a reputation of being a female killer, but he was just such a beautiful-looking croc that Mike couldn't pass him up. So he brought him here and he lived by himself for about a year and then we put him in the pen with two females and he just beat the crap out of them. We said, 'You're done,' so he was sold." I ask whether it matters if a croc is aggressive toward people or whether it's just the intraspecies bullying that affects these decisions. "It's usually about the females. If they're aggressive to me, I don't care. I just stay away from them." The male croc habit of "beating up" females is in fact a motivating factor in Bindara's efforts to develop methods for artificial insemination. This has been done with some success in alligators and other crocodilians, but it's still a work in progress with saltwater crocodiles. Collecting and storing the semen has proved to be relatively easy. Electroejaculation has been trialed, but the semen tends to be contaminated with urates (crystals found in urine). This is why digital massage seems to be a preferable method. First, the crocodile is captured and restrained. This is done by looping a lasso over the croc's upper jaw, which compels him to roll over several times, obligingly binding his own snout. Once the bite threat is assuaged, the handlers wrap duct tape around his snout and sedate him. They put a sling under his thorax and forelimbs and then remove him from his pen and load him onto a transport container that straddles two supporting structures for the semen-collection procedure. With the crocodile suspended overhead, the digital masseur—the guy with the glove on—positions himself under the animal. He inserts a finger into the croc's cloaca and hooks out the penis.[34] The finger is then inserted past the base of the penis, and the region around the urodeum is stroked down the proximal length of the shaft. This allows the white, viscous semen to flow down and drip into a collection tube. That semen can be collected again from a croc after thirteen days suggests that the trauma of capture and restraint is not an inhibition to this method, provided the croc is accustomed to a captive environment and constant handling.[35] In fact, the biggest inhibition to this procedure is the crocodile's ability to avoid capture. Once he's taped up and suspended, he drips semen like an espresso machine.

When Robbie presented a paper on the farm's progress in artificial insemination and selection for heritable traits, a local newspaper put out the story that Bindara was creating a breed of "super crocodiles."[36] It was a classic scare story, but it overstated the progress that has been made. Semen collection and storage has proved relatively easy, but it's only half the battle.[37] The procedure that has yet to be perfected is the insemination. The challenge here is getting the timing right. Research needs to be done on this aspect of the process, but there isn't enough funding available to move forward. If and when it does happen, artificial insemination will make for a very different crocodile farm. Rather than maintain a large number of breeding males—each with only a few females—it will be possible for an entire population of females to be impregnated with the semen from only a few stud males. It will even be possible to have no male breeders at all, because semen can be purchased from somewhere else. It will give croc farmers a lot more control over the genetic stock and eliminate the uncertainties that stem from the fickleness of male crocodiles in accepting mates.

Alf Hornborg describes how modernity implies a "shift from concrete to abstract reference points."[38] He's talking about how the conception of self is reconfigured so that specific places and people are detachable from a person's identity. But he also says that this objectification of self is in lockstep with the objectification of social relationships, and how both things tend toward alienation and fetishism. I've already described how conservation has traditionally served late modernity more faithfully than it has served ecological systems, how it has separated and abstracted things to a point where sets of numbers matter more than individuals. Here lies the irony in crocodile farming: what was born of a concern for conservation has, through its investment in particular animals, aggregated the abstract and concretized it to fit seamlessly with the paradigm of unmaking. Beyond factors like slow growth rate, diminishing clutch sizes, and exorbitant invoices for chicken carcasses, crocodile farming is up against late modernity itself. By directly engaging with crocodiles, in all of their idiosyncrasies, crocodile farmers are forced to engage with concreteness within a socioeconomic context that demands abstraction. Perhaps this works for a tourism business; on a farm tour, it's far more entertaining to speak the names of crocodiles, tell their histories, and explain away their recalcitrance by reference to their relations with other crocs, people, the water, and the weather. But crocodile farming is primarily about producing quality skins, and this is made more difficult when particular crocodiles do particular things—particularly fighting. Unlike a shed full of chickens, a farm full of crocs demands attention to the details of social relations. Farmers need to know who is bashing up whom, and how often. They need to think about what makes a particular crocodile behave in a certain way and work toward changing conditions to suit the croc. Farmers are dragged into the murky social worlds of crocodiles and made to accommodate their social needs. This is a direct challenge to unmaking and one that will not go unanswered. As much as farmers intend to approximate the wild, the crocodiles compel them to unmake their animals—to separate males from males, eggs from nests, eggs and hatchlings from cold and predators; to vaccinate and inoculate; to isolate. These are the kinds of separations that define domestication. The external fence keeps the crocodiles in, but the internal fencing is also extensive. Crocodiles are separated from one another and from predators, parasites, fungi, and the weather. They must be in order for the farm to persist. And artificial insemination is a logical extension of what must inevitably occur in this domestication context; it is a solution to a problem of social, ecological complexity. When artificial insemination comes to fruition, a brute of a male with otherwise desirable traits need no longer be sold off because of his habit of killing females. He can be kept in isolation, digitally massaged, and mated remotely with females a fraction of his size—females who would not stand a chance of survival were they locked up with him in a pen. Meanwhile, the females can be kept in isolation from other crocs and never suffer the ordeal of cohabiting with an aggressive male. There will be no problems with mating crocodiles because there will be no mating. The crocodile farm of the future will indeed "keep it PG."

5.

Emus

Flightless crim always dreaming of escape
Our helpless land a pan banged against summer bars
Whose true salute is money
—ERIC BEACH, "Emus Out of Genoa"

Emus don't fly. Never mind that they're six feet tall and thrive in the harshest environments, or that they lay emerald eggs that are as hard as stones; it is their flightlessness that fascinates both Western science and Aboriginal ontologies.[1] The question that plagued science for some time was whether the ancestors of emus ever flew at all. Emus belong to a family of birds called ratites that also includes cassowaries, ostriches, rheas, and kiwis, as well as the now extinct moas and elephant birds who weren't sufficiently fleet of foot to evade annihilation at the hands of men. One thing that links these birds is their collective groundedness. They are so obviously birds—they have beaks, claws, feathers, and vestiges of wings—yet they distinguish themselves from other birds by not flying. This shared family trait leads to the logical assumption that they are all descended from a common flightless ancestor. This assumption is halfway correct in that there was indeed a single species living some ninety million years ago from whom all of these flightless derivatives are descended. But this in turn creates a problem. Emus are indigenous to Australia, ostriches to Africa, rheas to South America, kiwis to New Zealand, and cassowaries to Australasia. These places were once part of the supercontinent Gondwanaland, which geology tells us split apart some 150 million years ago—sixty million years *prior* to the speciation of the various ratites. So if those ancestral birds were flightless, how did they cross over so much ocean to create the diversity of species we find today? It's a perplexing question, and one that has led to a lot of hypothesizing about various migration routes, including across Antarctica when the continent was not so frozen as to require the use of snowmobiles.[2] But advances in molecular biology tell an even more bizarre story: the mildly volant tinamous of the Americas are now included in the ratite family. They are in fact more closely related to rheas than rheas are to ostriches and emus. This inclusion of tinamous suggests that the ancestors of the contemporary ratites flew to their respective continents, and all but the tinamous independently lost the ability to fly.[3] This convergent loss of flight tells us that their common ancestor was probably not a very accomplished flyer, a theory supported by the fact that tinamous are rubbish flyers. Now, the question is, if the ancestors of emus were also rubbish flyers, how did they make it across so much open ocean? Perhaps their ability as swimmers is underrated.

Underrating emus' swimming ability is what anthropologist Kenneth Maddock did when he recorded a taxonomy from the Dalabon people of southern Arnhem Land.[4] According to Maddock, the Dalabon assigned all nonhuman animals to one of three taxonomic groups. One group, *djenj*, includes all species of fish. Another, *gunj*, includes the large marsupials. Last, there is *manj*, which comprises other species such as lizards, possums, bees, and birds, including emus. Maddock saw an anomaly in this taxonomy. The animals in the first two categories move in only one element. Fish swim in water; large marsupials move about on land. But the animals in the third category move in two elements. Lizards swim in the water and crawl on the ground, possums walk on the ground and climb trees, birds fly in the air and walk on the ground. But here is the anomaly: according to Maddock, emus only walk on land. Yet they are so obviously birds that they present a problem for the Dalabon: these people need to reconcile emus' relatedness to other birds with their categorical misplacement among other animals that dwell in two elements. This problem, Maddock argues, is solved by means of a dreaming story. In the story, the emu character is always withholding food from the other birds. One day, they happen to kill a kangaroo, and they trick Emu into searching for firewood for the cooking process. They keep sending her farther away to fetch more wood, so that she misses out on the kangaroo meat. Emu becomes disgruntled, and the birds launch themselves skyward with a kangaroo tail for a prize, leaving Emu on the ground. They call out to the dejected emu, "We're leaving you by yourself now!"[5] Maddock provides two analyses that seek to connect the myth with the taxonomic anomaly. In one analysis, it is the emu's lack of sociality that leads the other birds to exclude her. In the other, it is the way she stands apart from the others; at the beginning of the story Emu is presented as the leader of the birds. In both analyses, Emu's characteristics antedate her separation from the other birds and therefore solve the taxonomic problem of Emu's being included in the category of creatures that move in two elements.

But there is one critical flaw in Maddock's analysis: emus swim. Whether in coastal waters or inland lakes and rivers, emus are quite fond of paddling in the water. In fact, it's a sight to behold—a mob of emus crossing a river with only the tops of their backs and their long necks above the water; it looks like a raft of sea monsters.[6] Perhaps the high amount of body fat in emus makes for excellent buoyancy. So really there is no anomaly to be resolved. Emus move in two elements just like their bird cousins, only one of them is water instead of air. While I admit I'm being a little rough on Maddock—he was writing in a time when structuralism was in high fashion and data tended to serve theory—I think my criticism raises an important point: in trying to understand how the world speaks to our informants, we should pay attention to that world and its complexity rather than abstract things and confine them to categories. Maddock also notes that in one myth Emu swallows a stone that confers on her descendants the ability to lay eggs. Again using a structuralist analysis, he explains how stones are feminine, and that therefore they make sense in an egg-laying context. Had he witnessed the skinning and preparation of an emu for cooking, however, he would have seen the actual stones (known as gastroliths) that emus swallow and retain in their gizzards to help break down fibrous foods.[7]

For many Aboriginal Australians, stories are more than ways of sorting out categories. They connect strands of the world, reiterating ecologies and binding together humans, other animals, plants, rivers, hills, trees, and the cosmos. If you're ever in Sydney, it's worth making a trip to Ku-ring-gai Chase National Park. There you can follow walking tracks among the banksia and scribbly gums where the forest touches the sea, where creeks made amber from tannins feed placid inlets lined with mangroves. This is the homeland of the Guringai people, whose place in history was a little too dangerously close to the early colonial encounter.[8] Within two years of the arrival of the first colonists, most of the Guringai were wiped out by smallpox; the remainder were killed off or absorbed into settler society as the colony expanded.

But the Guringai left a legacy that will endure with the sandstone. They made hundreds of engravings in the rocks of what is now Ku-ring-gai Chase National Park. The National Parks and Wildlife Service has created a heritage walk along which you can see many of these engravings among 350 art sites. But there is one site in particular that lies on a little used track in a different part of the park that is very much worth visiting. At this site, there is an engraving of an emu carved into a flattish stone slab. It's a curious engraving in that the emu looks like he's been stretched and laid out across the stone. It doesn't seem to represent a typical emu. Among non-Aboriginal people, it was Sydney academic Hugh Cairns who eventually made sense of the emu's weird posture. Cairns made a connection between the engraving at Ku-ring-gai and a widely held Aboriginal view of the cosmos. The engraving represents a part of the night sky.[9] It's a bit hard to see so close to Sydney, but farther out in the bush where the night air is clear and the urban lights don't shut out the stars, you can see the dark emu amid the stars of the Milky Way. At first he's difficult to discern because he's not composed of stars. Rather, the emu in the sky takes up the negative space between stars, the shadowy cloud of interstellar dust where new stars are being born. Once you see him, though, you will never see the night sky in the same way again. His head is the Coalsack Nebula just below the Southern Cross. His neck and body extend below, just as in the engraving, all the way down to his claws. There are even two stubs of wings that are astonishingly emulike. This sky emu is known to Aboriginal groups across Australia, from the Sydney Basin in the east to the Kimberley and the Central Desert; from Tanami and the Murchison Ranges in the Northern Territory to west-central Victoria in the south.[10] There are countless dreaming stories explaining how the emu ended up there, stories that connect the cosmos to emus and ecologies.[11] One such story, from the Wangkumara people of northwest New South Wales, tells of a giant emulike creature, Ngindyal, who terrorized people across the region. She was so big that her eye could be seen in the sky as a star. Eventually betrayed by Crow, Ngindyal was killed by two brothers—ancestors of the Wangkumara—and her body came to occupy the dark sky below the Coalsack Nebula, known locally as Maringa Bambu.[12]

While the meanings implicit in the emu engraving are lost with the Guringai people, we can get some idea of how it held meaning for them by comparing their sky emu with stories of the Kamilaroi of northwest New South Wales, from whom we do have some sources. While the head of the emu can be seen all year long, his body and legs dip below the horizon to varying degrees depending on the earth's tilt in relation to the sun. When the emu's legs are visible, it is mating time, when females chase males. As the legs disappear below the horizon in May through June, the emu appears to be sitting. This is the time when males incubate eggs; thus it's time to go collecting emu eggs. It's also the time to conduct male initiation ceremonies called bora.[13] This makes sense because both the incubation of eggs by male emus and male initiation among the Kamilaroi denote the same things: young males taking on responsibility and older males nurturing initiates. In late winter, the emu's neck becomes indistinct and his body represents an egg. This is the time when chicks hatch and are cared for by their fathers. In the late spring, the emu sits low on the horizon in the evening sky, which represents the emu sitting in a water hole. This corresponds to the time when the water holes are full. Then, in summer, the emu becomes almost invisible, denoting the time when emus have left the water holes and are scattered throughout the landscape.[14]

On the Maburrinj Estate in Arnhem Land, there is another emu laid out on the ground, only this one does not look like the emu in the sky. It is represented in relief as a stone arrangement on an important walking route of the Bininj people of western Arnhem Land. The name of the stone arrangement doesn't quite evoke the majesty of the Kamilaroi sky emu. It is locally known as Kurdukadji Dedjbarl Kahrrmeng, or "Emu slipped on her arse."[15] While there may be wry humor in the naming, the site is still taken seriously. It's an emu increase site, where visitors strike the stones with branches and call out to their ancestors, asking that they be provided with plenty of emus. A photograph of the site appears in a book called *Something About Emus*, which documents the importance of emus to the Bininj. The book emerged out of a project by a Bininj ecologist, Dean Yibarbuk, aimed at recording his people's stories about emus. It's also a linguistic project in which the language, Bininj Kunwok, is used as the primary text, with accompanying English translations. In the first section, Bininj people talk with one another about emus. These people are deeply immersed in knowledge of emus; they know intimate details about their habits and movements, about their predators and predilections, about what it is like to be an emu. A lot of the discussion is about what emus like to eat: green plum, black plum, quinine tree, red leea fruit, holly-leaved pea flower, blue tongue fruit, kangaroo blood berries, bush carrot, wattle flowers, grevillea flowers, water weed, algae, and, of course, emu apple. They also eat charcoal, possibly to aid digestion. Emus have catholic tastes.[16] A deep knowledge of emus' diets is an effective tool for hunting. Mick Kubarkku describes how he used to hunt emus by sitting in the fork of a black plum tree. He'd find a tree with emu poo below, climb the tree, and wait for the emu to return. The emu would come to eat the black plums scattered on the ground and he'd spear her from above. It was an effective ambush because the emu was looking at the ground. Nowadays, he uses a firearm. Meanwhile, modes of hunting emus at ground level capitalized on their idiosyncrasies. Emus are notoriously curious creatures and their curiosity can be exploited. Jack Djandomerr, another Bininj man, tells how he attracts emus: "You get an axe and strike the ground *dub dub dub* and it will run up close to see what you are doing and you keep hitting until it gets close and then you spear or shoot it."[17] People also hunt emus by smearing their bodies with white ochre and hiding behind branches that they hold in front of themselves. The camouflage allows them to get close enough to spear the birds.[18]

Among the nonhuman predators of emus are dingoes, snakes, and black-breasted buzzards. None of these animals kill adult birds, but dingoes and snakes kill chicks. Black-breasted buzzards eat the eggs; Jack Nawilil, also quoted in *Something About Emus*, describes their method: "The black breasted buzzard gets a stick. It runs up with a stick in its claw and breaks open the egg. Then it eats."[19] These people are also well versed in emus' swimming habits. Peter Nabarlambarl describes how they go for a swim in the morning and again after their afternoon rest.[20] And they go into great detail about emus' nesting habits—the way they sprinkle eggs with water and the tree species from which the emus strip bark for the nests. The eggs themselves are made sacred by these people. Elders incorporate them into what is called the Mardayin ceremony, making them *duyu*, taboo, forbidden to children. And various body parts of emus have particular names, with rules for their distribution. For many Bininj clans, emus are ancestral figures within both Yirritja and Dhuwa moieties, so, yes, they are related to stingless bees.[21] Emus and humans share a common, humanlike ancestor in the dreaming, and there are stories describing how Emu became a bird. Thus when an emu is killed, a member of an emu clan is made responsible for distributing the meat. They are effectively sharing a family member.[22] *Something About Emus* also contains an account, in Bininj Kunwok, by the anthropologist John Altman, who witnessed the return of an exhausted but elated young hunter, Joshua, carrying a fifty-five-kilogram emu. Joshua's uncle was made responsible for the cooking and distribution, so it is likely that he was of an emu clan while Joshua was not. We know that Joshua was not because he killed the emu; this is forbidden to members of an emu clan, and had one been present, Joshua would have needed his or her permission. George Djandomerr, another Bininj, also emphasizes the relatedness of people and emus, describing how emus are considered nonhuman persons. They are deeply familiar with the landscape and remember where they found food two or three years before. They follow the seasons in their foraging, moving through the region as different foods become available. Djandomerr says, "Bininj Wanjki"—they are just like people.[23]

Emus are consummate travelers.[24] The Bininj people describe how emus travel from place to place in search of food. When a particular area of country is burned, the emus disappear, but they return months later when the trees have regenerated and bear fruit. Emus don't maintain territories, and only when incubating eggs do they stay more than one night in any given camp. "They sleep anywhere and they set off and turn up at another place. Emus, they wander around all over."[25] Unsurprisingly, scientific inquiry into emu movements backs this up. In 1969, researchers from the Commonwealth Scientific and Industrial Research Organisation captured and banded 154 emus in central Western Australia in order to track their movements and determine what constrained them.[26] They had remarkable results with respect to how far some of these emus traveled. Band number 108 was applied to an emu captured on the eastern side of rabbit-proof fence number 1, near Wiluna. The band was recovered near Lyndon some eight months later, 539 kilometers away as the crow flies. Of course, emus don't fly, so, assuming a meandering path, this bird would have covered considerably more distance. She also managed to get through, or over, the rabbit-proof fence. A lot of emus traveled at a leisurely pace of only a few kilometers a day, but in some cases they moved at a rate of more than thirteen kilometers a day, again measured as a straight line. Interestingly, they might have traveled greater distances if not for the rabbit-proof fences. The emus were clustered along these fences when first captured for banding, and in most cases recovery was also made along fence lines. They seem to have habitual, seasonal migratory routes that are now inhibited by the fencing. Environmental historian Libby Robin writes admiringly of the way in which emus adapt to a "boom-and-bust" landscape. Wherever resources become available, emus appear as if from nowhere and rapidly increase in number. As resources diminish, they go somewhere else. Robin suggests that colonial Australians build economies more aligned with Australian ecologies by paying attention to the adaptive rhythms of emus' movements across ephemeral landscapes.[27] Unfortunately for many emus, colonial Australians have been less interested in learning from them than in exploiting or eradicating them.

Probably the most famous example of attempted eradication is the Emu War of 1932.[28] After the First World War, returned servicemen and British veterans were given parcels of agricultural land by the West Australian government. They enjoyed some good seasons throughout the 1920s until the Great Depression hit. The government encouraged the farmers to abandon mixed farming and go into intensive wheat production, which many did. But soon thereafter the price of wheat plummeted. Wheat became heavily politicized, and farmers threatened to withhold the wheat harvest of 1932 in protest. Enter the emus. In the midst of the wheat crisis, an estimated twenty thousand of them descended on the Campion and Walgoolan wheat-growing districts and began consuming the wheat crop. Farmers demanded that the government intervene, and the government came up with a plan to eradicate emus and at the same time provide valuable training for military personnel. Major G. P. W. Meredith of the Royal Australian Artillery and two machine gunners (with ten thousand rounds of ammunition) were given the task of eradicating the emus. Seeing the potential for propaganda in the campaign, the government also sent a cinematographer to document the events. As it turned out, both decisions were sorely misguided. The first engagement took place at Campion, where the machine gunners opened fire on forty or fifty birds, who ran for the cover of some trees. Major Meredith had expected to encounter masses of birds in open country, but the small groups of emus, who kept close to cover, made mass killing impossible. The men tried to ambush the emus at water sources but managed to kill only a few birds. When they finally found a large group of birds in open country, the machine gun jammed. Some soldier-settlers who accompanied the party opened fire with rifles, but they succeeded in killing only a few emus before the flock ran for cover. By November 8, Meredith's men had used twenty-five hundred rounds of ammunition to kill a reported two hundred birds. They noted the toughness of the emus—one farmer ran down an ostensibly healthy emu with his truck and found five rounds in her body.

Meanwhile, in the city, the operation became a political disaster. A question about cost and effectiveness was raised in the Australian Parliament and the prime minister had no answer. A New South Wales Labour politician asked wryly whether a medal should be struck to commemorate the campaign, to which a West Australian colleague replied that any medals should be given to the emus, who had "won every round so far."[29]

The following day, the defense minister ordered Meredith and his troops to withdraw. A month later there followed a second campaign that was equally ineffective. A tremendous amount of ammunition was used, and the Agricultural Board audaciously tried to recoup the cost from the farmers. One farmer responded that the board in fact owed him money for his time spent dispatching birds wounded by the machine gunners. It was a farce that cut straight to the weak underbelly of colonialism. It highlighted white Australians' awkwardness in engaging with the land and their hopeless reliance on traditional methods of colonization—militarism, exclusion, and eradication—in the face of a recalcitrant ecology.

After almost one hundred years of persecuting emus, colonial Australians decided instead to make them productive.[30] Somewhat ironically in light of the Emu War, emu farming began in the 1970s in Western Australia. The first farmed emus were caught in the wild and raised in fenced enclosures on primarily grain-based diets.[31] Their placid natures and tough constitutions adapt them well to an Australian farming environment, and their flightlessness makes fencing practicable. By the early 1990s, all Australian states had enacted legislation to allow farming of emus, and within a few years more than sixty thousand birds were being farmed across Australia.[32] By this time, emu farms existed in the United States as well, and the American Emu Association (AEA) had been established. Before Australia imposed a ban on exporting emus, many had been sent to the United States to be bred as novelties, but in the 1990s emu farming came to be seen as potentially lucrative; by the mid-1990s the AEA had more than six thousand members. Around this time, the Rural Industries Research and Development Corporation in Australia was funding projects aimed at improving production, and emus were becoming increasingly commoditized. One such project took a very clinical approach to food restriction and the timing of slaughter. A 1999 RIRDC research report found that profits could be optimized by means of a dual-cohort model. Early-hatched birds were to be fully fed until they were slaughtered, while late-hatching birds were to be fed a restricted diet until eight weeks prior to slaughter, whereupon they were to be given a high-energy "finishing ration" that caused them to gain 1–1.5 kilograms of weight per week. Both cohorts, ranging in age from fifty-six to seventy-two weeks, were then to be slaughtered during the period September to May. The report recognized the limitations of this model but found it to be the most profitable method of raising emus: "Although this will confine the supply of fresh product from September to May," it said, "each year the killing of birds outside this age range has a considerable cost penalty."[33] It goes without saying that the position of the Coalsack Nebula in relation to the horizon has no bearing on the economics of emu farming. The commoditization of emus led to high demand for breeder birds, which created a market in which, somewhat like stingless bee hives, the birds themselves were more valuable than the products derived from them. It was a classic recipe for a financial bubble, and as breeders bred, the birds lost value and the bubble inevitably burst.[34] Today, the membership of the AEA can be counted in the hundreds. Still, emu farming persists in Australia and overseas, and the farming model is based on the products derived rather than on the birds themselves. As of 2014, the population of farmed emus in India was 1.4 million across seven states. The separation of emus from their natural ecologies has thus reached global proportions, and while they may never reach the scale of mass production of their hyperproductive cousins, chickens, they are being adapted to a farming environment with almost the same degree of vigor.[35]

Emus are seen as viable poultry birds because their bodies yield multiple potential products. Their body fat has therapeutic properties, their meat tastes somewhat like beef, their skins can be tanned into leather, their feathers can be used as decorations, their eggs can be eaten or emptied and used ornamentally, and their claws can be polished and used in jewelry. Of all of these uses, emu oil has received the most attention.[36] An adult emu weighing forty-five kilograms carries about ten kilograms of body fat, which can be translated into seven or eight liters of thick yellow oil.[37] According to research, the low triglyceride content of the oil lends it anti-inflammatory effects. Aboriginal people have long known this and reportedly have used the oil for healing wounds, pain relief, and treating musculoskeletal disorders. When used with a transdermal agent such as eucalyptus oil or cineole, the healing effects are said to be enhanced, although the anti-inflammatory effects are in turn thought to be dependent on the diets of the emus.[38] Wild emus provide the most efficacious oil, followed by "naturally" farmed emus, with intensively farmed emus providing the least efficacious oil.[39] Applied topically, emu oil has also been shown to have greater anti-arthritic effects than rhea oil, fish oil, olive oil, flaxseed oil, evening primrose oil, or lard from pigs. As a prophylactic, it also suppresses the development of arthritis. Where it's applied topically, emu oil has been shown to cause little to no irritation, unlike transdermally active drugs. In one experiment, emu oil was shown to reduce inflammation of the alimentary tract following chemotherapy. Given orally to rats, the oil was shown to improve mucosal architecture in the intestines during recovery from cancer treatment. Other studies suggest that it also suppresses bone loss after chemotherapy. In addition, it has hypocholesterolemic properties that are said to reduce both total cholesterol and the ratio of total cholesterol to high-density lipoprotein. In other words, it may be effective in treating people with high levels of "bad" cholesterol. One U.S. patent claims that it stimulates melanogenesis (skin growth), reduces wrinkles, and rejuvenates aged and damaged skin. This patent also claims that it can treat male pattern baldness and alopecia and can be used as an effective treatment for burns and psoriasis. It has been promoted as a transdermal enhancer and even touted as an insect repellent. There are also claims that it has beneficial effects in people with diabetes. Proponents point out that, unlike petroleum-based products, emu oil needs little refining, is easily metabolized, presents a low health hazard, and comes from a sustainable source.[40] Stop pushing at the back, please, ladies and gentlemen, and only two bottles per customer.

It's been my good fortune to encounter emus in all manner of places. Once, while I was mountain-biking around Wilsons Promontory in Victoria, a trio of emus appeared in front of me. It so happened that the path was lined with coastal scrub on both sides, so the emus had no choice but to jog ahead of me for what must have been a kilometer, while I had to slow up so as to not startle them further. It was awkward. Another time, I camped on the grassy lakeshore of Copeton Waters surrounded by emus and kangaroos. That was idyllic. Most of my encounters, though, have been at the sides of highways. I've seen them grazing with sheep on stubbled fields, standing in the shade of solitary trees, panting to stave off the midday heat, and stranded and starving in the narrow ecozones that fencing makes of highways. So it's not at all unusual to be standing among a hundred emus in a small paddock while deafening B-double trucks roar past on the A3. This is Emu Bliss, an emu farm in South East Queensland. It looks like a regular farm; there are fenced paddocks of green grass with assorted bits of rusting farm machinery, corrugated iron sheds of various sizes, feed troughs, water tanks, and decorative trees lining the drive up to a Queenslander-style farmhouse. It's a little unusual, though, as farms go, not just because of the giant birds but because it's open to the public. There is a shop where you can buy emu oil, emu eggs, emu stationery, and plush toys that look not entirely unlike emus. Today there's a large marquee in front of the farmhouse with rows of white plastic chairs lined up under its shelter. The owners are holding information sessions for people interested in the therapeutic effects of emu oil. I find myself a seat under the marquee and take stock of the audience. They are mostly elderly folks in varying states of discomfort and disability. Shane, the owner-operator of the farm, arrives and introduces himself. If there was a type specimen for an Australian farmer, Shane would fit the bill. Tall, red-skinned, and rotund, he sports an Akubra hat and R. M. Williams boots. The only thing that distinguishes him is his bright red shirt. As they say, if you want to sell something, paint it red. Shane begins the session with the story of how he came to emu farming through using the oil. Some twenty years ago he had shoulder and knee problems, and a friend suggested that he put some emu oil on the affected joints. He tried it and was amazed at the results. So impressed was he that he did some research into emu oil and within six months had established Emu Bliss. Despite the red shirt, Shane is no snake-oil salesman. He doesn't use hyperbole, nor does he patronize his audience. But his passion for emu oil is plain to see. He tells us how he collects data from people who buy the oil. They record their blood sugar levels and these data go into a database. He worked with a medical researcher from a hospital in Brisbane on a report in the journal *Inflammopharmacology*; the report was crucial to getting emu oil approved as a medicine in 2002. He discusses the claims that it treats inflammation and atherosclerosis and tells us that in capsule form it can even treat ADHD. At the conclusion of the talk I'm not exactly jumping over the chairs to get myself a bottle of emu oil, but then I don't have any of the conditions it's supposed to treat. He hasn't mentioned hair loss. Still, I'm convinced of one thing: Shane firmly believes in his product.

After the talk, Shane takes me for a look around the farm. Our first stop is the shed where chicks are reared. This is what he usually shows the tourists because, well, cute baby birds. Shane tells me to wait in front of the exercise yard while he herds the chicks out into the yard. The week-old chicks pour through the door and crowd into the open area. Emu chicks are quite stunning. With their teardrop bodies, stubby beaks, and bold black-and-white stripes, they look like the hybrid offspring of Eddie van Halen and a dipping bird. But they're also adorable and vulnerable-looking, and I can see how they bring out the protective, nurturing side of male emus.[41] The camouflage and the way they flock together tells me that emu chicks are nervous about predators, as indeed they should be. If not for the electric tape that Shane has set up around the shed, the local foxes would devastate his flock. They're also frightened of me standing outside the exercise yard, so they try to outmaneuver Shane and get back inside the shed. He's not given to stressing them out just for the sake of a visitor, so he lets them back in. "At this age," he tells me, "you need to make sure that you interact with them as little as possible. This is because they imprint on you and it causes all sorts of problems later on." What sorts of problems? "As adults, they think you're a potential mate, and not only do they follow you around; they try to stop you from getting to the other birds. It makes it really hard to manage the other birds. That's why we try not to pick up the chicks. We only go in there to do the food and water." Shane also employs slightly older chicks to teach the hatchlings how to use the feeders. "When we get newborns, we always take a small one from the last batch and put him in the shed as a teacher. It gets them eating a day or two earlier by having a little teacher in there. We don't have to do it, but we want to get our birds as big and fat as we possibly can." I ponder the unmaking of these emu chicks, separated from predators, parents, and people. They exist in a world of other emu chicks, water bowls, feeders, and not much else.

On the other side of the shed, Shane shows me two paddocks where the older chicks are kept. In the first one there are about a hundred birds at three months old. They wear the vestiges of their baby stripes but are tall enough to look like miniature emus. Just like their younger cousins, these chicks are in antipredator mode, and as we approach the fence they make a beeline for the other side of the paddock. They are the very opposite of what I would imagine an imprinted bird to be like, so the lack of handling seems to have worked. But the contrast with the adjacent paddock is incredible. Next door there are about fifty six-month-old emus who have gathered at the fence, falling over one another to get as close as possible to us humans. Shane tells me that these birds are in no way imprinted. Rather, at about five months, their chickish timidity is replaced by adult-emu curiosity that could only be greater if they were holding magnifying glasses. This is the notorious curiosity that the Bininj hunters capitalized on. Shane tells me how they're especially drawn to shiny objects. "Emus in the zoos will go for people's diamonds and earrings. I've had the odd phone call from a zoo as somebody's earring got swallowed or something like that. . . . I tell 'em just wait a couple hours and it'll come out the other end." Does this make it hard in terms of farming? I ask. "Oh yeah, if we're fencing, we've gotta be very careful that they don't eat our screws and anything else that we use. We gotta be very careful of the little offcuts of wire and stuff like that because they just eat it. And they try to help when we're fencing. If you're doing fencing in a pen, they'll come up and try to help us." I notice that there are a few shelters but the emus aren't exactly crowding around competing for shade. They don't seem too concerned about the sun. "We do put the water underneath a shelter. Not so much that the emus want shelter—they don't need it. They can have shelters, they can have trees, and they're still gonna be out in the sun. They're all insulated. Doesn't matter whether it's hot, cold, dry, wet. They're adapted to all sorts of climates in Australia. We put the water under the shelters just to encourage the birds when they wanna have something to drink, but after they go into the shelter and have a drink they'll just go out again anyway."

Back at the farmhouse, I'm interested in the caravan parked out front. There's a cartoon emu painted on the side with a caption: *I can't fly but I'll tell you true, I can run the pants off all of you.* It's a paraphrase of the famous song "Old Man Emu" by John Williamson. Shane explains how he takes the caravan to exhibitions and agricultural shows to promote his emu oil. There's a separate compartment at the back. This, he says, "is where we used to keep Doc when we took him along." Shane shows me a newspaper clipping with a picture of himself and an emu standing alongside. "Doc was hatched right at the end of the season. He came out with a big belly and very tiny legs, and his feathers touching the ground. As he grew up, his legs started to bend out a bit. Normally, when they start bending they'll actually go out to the side and then they can't walk, but Doc's only bent a little way and then he was okay. When he was six months old, I took him to a show and he never pecked anybody. That was the only emu I've ever seen that won't peck at people, so I took him to shows for about five years. One time I was walking with him down the main street of Gladstone, and you should have seen the looks that we got." Doc is no longer with us, but his legacy remains. He is the emu who appears on the packaging for the farm's emu oil products.

The next day, I follow Shane through the back roads of southern Queensland to his second farm. This is larger than the roadside farm and has no residence. Amid a wider landscape of pasture, cattle, and scattered eucalypts, the high fencing and giant birds of an emu farm are rather conspicuous. Shane explains that the mesh fencing here is of a smaller gauge compared to the other farm. "What we learned the hard way is if the gaps in the wire are too large, they push them open and try to get through." I'm reminded of the crocodiles who escape their pens but forego their chance at freedom. What would happen if they got out? "They'd go." They wouldn't hang around? "When they're fenced in, they find a particular place for themselves in the paddock, but without the fence they'd run." At the top of the driveway,

Shane's son Brendon is working on some concrete footings. They're building an onsite slaughterhouse and processing plant that will save having to transport emus to the other farm. The business is expanding, but there is a limit. Shane says that one thousand birds are about as many as he can handle before it becomes too much work. Still, that's probably more than the Royal Australian Artillery could manage.

We climb aboard the tractor with a trailer load of feed and drive through a gate into one of the paddocks. Shane is at the controls, Brendon is clinging to the ladder on the feed trailer, and I'm riding shotgun on the tractor's side step. I can either dangle one foot in front of the massive, spinning tire treads or bend one leg over the wheel arch. Both positions are uncomfortable, and I suspect the side step was designed to dissuade people from riding on it, or else to punish them for doing so by dragging them under the rear tires. Shane notices that some of the feed bins are empty, and Brendon mentions that the emus have been looking hungry lately. "That's a good thing. The more you feed them, the more fat they can put on." An emu's gut is comparatively short; in a farming environment, food can pass through so quickly that after it's excreted another emu will eat it. To try to stop this, they add hay to the feed to help it stay in the gut a little longer and get broken down. Since fat is the primary product of the farm, everything is aimed at maximizing body fat at the time of slaughter. "We're always doing research and trying to get the maximum amount of fat," Shane says, "which gives us the maximum power in our oil when we make it. The main part of our business is getting these ten fatty acids to be as powerful as possible, because the acids dissolve cholesterol and fat stuck to artery walls." This is why there are three feeders and three water troughs in each paddock: so the emus need not travel too far to feed, or cross over too many other emus' territories and cause squabbles.

We stop at a feeder and Brendon alights so that he can lower a spout from the trailer and fill the huge feed bin. I'm struck by the different levels of interest shown by the emus. Some are right up at the feeder as soon as it's filled, while others hang around at the other side of the paddock. Some follow the tractor, while others walk ahead of us. But there is certainly no crowding or squabbling over food. According to Shane, there is very little aggression among the emus. If he sees birds being picked on by other birds, he removes them to an "exiles" paddock. These birds may well breed, but the eggs will be collected, drilled, and emptied rather than hatched because he considers them misfits. Meanwhile, overly aggressive birds simply find a mate and establish a territory within the paddock where they keep to themselves. If a bird follows Shane or Brendon around, she is moved to the exiles' paddock as well. "If we're ever walking around the paddock—collecting eggs, for instance—there are some birds that, maybe because they've been handled too much when they're young, will follow us around the paddock. What that means is that they're then going into other pairs' territory. Then that bird that's following us wants to fight the others, because they don't want those birds near us. They're trying to keep other birds away from us because they think we're their partner. So those birds that follow us, we put them in the paddock with the picked-on birds. That way we have a peaceful farm and not getting birds that are wandering around into somebody else's territory." As we drive through the paddock, I notice a contrast created by the fencing. Outside the fence the grass is as high as nine strands of mesh. But inside there is a bare strip of earth along each fence line where the emus have walked back and forth, wearing a groove into the ground while they dream of travels beyond the wire mesh. We stop beside the fence and Shane indicates two shiny green eggs on the ground. He asks if I could hop down to collect them, and I oblige. Handing them over, I ask what will happen to these eggs. "In the middle of the season, we would take them back to the incubator, but it's late season now. These ones will be emptied and sold as decorations." I ask what's wrong with eggs that are laid late in the season. "By then," Shane tells me, "the females are running out of fat, so those eggs may not be as good and such. Doc was a chick right at the end. He was a lovely bird, but he had all those physical problems."

When you're collecting eggs in midseason for incubation, I ask, is there a selection process? Shane explains that he looks first at the size of the egg and rules out not only the smallest eggs but also the largest—because a large chick might have too big a belly, which can put too much pressure on his legs. "We always look for the middle of the road, and anything from, say, a five-hundred- to a six-hundred-gram egg" is ideal. Shane also examines the shape of each egg, looking for those with a larger circumference at one end, smaller at the other. "If it's sort of narrow on each end," he says, "the chick might be a bit squashed into that sort of environment." From roughly 650 birds, he harvests about three hundred eggs a week at the height of egg season, he says, "so we can afford to be picky." Does it matter which birds lay the eggs? "We do look at our breeding birds," Shane tells me. "If we've got birds that are, for whatever reason, being picked on, we separate those; we'd always try to stay away from incubating eggs that are from birds that don't seem to fit in."

Shane stops the tractor again so that Brendon can fill another feeder. He indicates another egg by the fence on the other side of the paddock, so again I dismount and collect the egg—only this time I have to do the collecting among two dozen emus. The birds move out of my way as I get closer to the fence, and I pick up the egg, which is not emerald green but white. Apparently, if a female is disturbed while laying an egg, she will draw it back in and lay it at a later time. After the second attempt, it will be white. When I turn to carry my bounty back to the tractor, I find a gathering of stern-looking emus blocking my way. I try to saunter through the flock as casually and inconspicuously as possible, and I'm suddenly reminded of a zombie movie in which the human protagonist smears himself with gore and tries to penetrate a blockade of zombies without being noticed. Inevitably, he is noticed, and so am I. The emus crowd around me, making it difficult to get back to the tractor. Shane has prepared me for this, instructing me never to run away from aggressive emus, because you can't outrun them. "As they run over the top of you, they rip you apart. Always stand your ground. I was taught to never go into an emu paddock until you are confident that you are going to stand up to that emu. Don't run away! It was drummed in my head: don't run away!" Surrounded by emus as tall as I am and with clawed feet like predatory dinosaurs, only the last two words of Shane's tutelage are making any sense. But there was more that he'd told me: "Just hold your hand up high and go to them. There are some that will continually come at you, but the same thing: hand up high, be assertive. You don't have to be aggressive and slap 'em around—just be assertive." Inspired by Shane's pep talk, I drop the attempt at being inconspicuous, and, with a flourish, put my hand up high in the air with it bent over like an emu's head for added effect. Thankfully, the emus contort themselves as if to curtsy and clear a space in front of me. I take my low-rent puppet show all the way back to the tractor and hand Shane the whitened egg. "Were you concerned about me back there?" I ask. "We do have the odd one or two birds that'll be aggressive," Shane replies, "but it's the end of the breeding season, so things are fairly calm. I was keeping an eye on you." I ask if he's ever been hurt by an emu. "Oh, yes, I've been raked lots of times. Same with Brendon—he had a really bad one that tore through the muscle and needed stitches." Shane shows me his forearm. "I had a really bad one here." I inspect Shane's arm and find only the faintest of scars. In answer to my unspoken question he replies, "Emu oil."

Alongside the B78 in New South Wales, I find another farm with emus that stands in marked contrast to Emu Bliss. I say "farm with emus" rather than "emu farm" because emus are not the primary product here. Pecans dominate this farm, and not just in economic terms. The trees are huge. But emus, sheep, cattle, and avocados also have a place here. The farmer, Barry Watson, bought the thirty-five-hectare riverside property just ten months before, and he's making changes. The farmhouse has been repainted and is getting new flooring, cattle have been introduced, and the avocado trees and emus are on the way out. Barry and his farm manager, Ross, take me across lush pasture that almost qualifies as parkland to where the emus are held. We pass by poincianas, jacarandas, and bushes made into topiaries by spoiled sheep, to a paddock with mesh fencing. The first thing I notice is the groove worn into the ground inside the fence, exactly like the one I saw at Emu Bliss, which bespeaks a frustrated desire to go somewhere, anywhere. The path leads to the corner of the paddock, where the smallest emu of the mob, Archie, stands watching traffic pass by on the main road. "Archie is probably the most photographed emu in Australia," Barry tells me. "He's always standing in that corner and tourists are always stopping there to take photos." Possibly Archie was handled a lot as a chick, because he's very fond of people. This makes him easy to manage, because he'll follow Barry or Ross to whichever paddock they want him in, but it also means he tries to mate with them. Ross gives me a demonstration of Archie's malleability. He walks up to the small male bird, puts one arm around his chest, and gently strokes Archie's head. It's a profound display of affection, but for some reason it disturbs me greatly to see an ancient Australian bird in the embrace of a man.

I ask Barry if the emus are difficult to manage. "The thing about emus," he says, "is that they won't round up. So if one gets out of the paddock it can be a bit hard getting it back in. Sometimes a bit of coaxing works. One strategy is to put one arm over the body and then guide it by the neck. They're also very curious, so we can use that. They like galvanized buckets, so we might use one of them and get the emu to follow." But their curiosity also works against their keepers in a farming context. "If you're doing work in the paddock, you'll turn around and find one's run off with one of your tools. And they especially like wiring. If they get the chance, they'll pull the wiring out from our cars or farm machinery. But other than that they're pretty low maintenance." Do they have any predators? "They aren't bothered by wild dogs like the sheep are. Though if the dogs cornered an emu, it might have some trouble. Most of the predation is on eggs. Goannas and snakes take the eggs. The eggs are so hard, I don't know how they break 'em open. Probably wild dogs would take the eggs as well." Barry explains that this is how he's getting out of emu farming. He considered selling the birds, but he expected he would need a permit for that and it would take more time than it's worth. So, instead, he's letting his emu population decline by attrition. Unlike at Emu Bliss, he isn't collecting and incubating eggs, and because he's neglecting this one aspect of emu farming, his birds are not reproducing: predators are getting all of the eggs. These emus are at an ecological dead end. They will live out their lives pacing fence lines and chasing buckets, and their matings will come to naught.

I want to know the history of this soon-not-to-be emu farm, so I track down the former owner, Terry Walker, who's retired and living on the coast. When I call, he and his wife are in their RV, on their way north to Queensland. He agrees to break up the drive in order to meet with me. Sitting on my front porch with a coffee, Terry reminisces about his days farming emus. He initially got into emu farming because of his concern about the number of wild emus in the Northern Rivers region of New South Wales. "They're very prolific breeders, but they have a lot of predators in the wild. Foxes, wild dogs, and goannas are right onto the eggs. So it's difficult to maintain a good number in any particular area. I don't know about inland Australia, but in the Northern Rivers area the numbers are in decline." Part of Terry's venture involved helping the National Parks and Wildlife Service train dogs to find emu nests in the wild, presumably to salvage the eggs and incubate and hatch them for wild release. "The handler would come out to the farm on a regular basis with the dog. She'd have the dog smell the nest, smell the eggs, and smell the feathers to be able to find those things in the wild." Terry's emus had firsthand experience of the kinds of predation that threaten their wild relatives. "We had goanna problems. They're right onto it and they know how to break an emu egg. They pick one up in their mouth and break it on another one. Or if there's only one egg, which is rare, they can easily break it on the ground." What about wild dogs and foxes? "Yeah, one day we had thirteen little chicks from a clutch and the next day we had none. And I saw the fox getting through the fence. There's all sorts of predators out there. Wedge-tailed eagles take chicks, foxes will, wild dogs will. In fact, an emu farmer I know lost all of his emus to wild dogs in a single night. He was really well set up with fencing but they got in and got the lot." I ask if he did anything to prevent chicks' being taken. "Yes. Where we had chicks, we put chicken wire up to a level that foxes couldn't jump.

And we put solar lights up with sensors. We had solar-powered electric fence around most of the property. But then you'd get one big flood and all your equipment's gone. I think we did quite well when I look back—they were well cared for." So Terry didn't constantly take eggs from the paddock to incubate. He took some to eat and, ironically, he took some in order to keep the emus from breeding too prolifically, but his approach was to let the emus breed in their own way and take steps to keep predators out. He didn't produce oil, eggs, or meat for sale; he farmed for the love of emus—but even so, his emotional investment compelled him to separate his emus from the wider world of predators and pests.

In terms of farming, Libby Robin's suggestion that settler Australians might model a better economy by emulating emus' ways of living in a boom-and-bust environment is ironic. Rather than adapt to emu ecologies, we expect them to adapt to ours. The way we conceptualize land—as something that can be broken down and sold in divisible units—is anathema to the itinerants of the bush. I've seen the grooves worn into the ground along fence lines and the vast concentrations of emus that accumulate at the border because of a continuous length of wire mesh. Something in the nature of emus compels them to keep it moving. But the farming, and by extension the domestication, of emus requires that emus be severed from their traditional ephemeral ecologies. Modern farming demands ownership of individual birds on acreages that are small enough to belong to individual farmers but too small for emus. Fencing makes almost perfect sense in such a world, at least to the farmers. It confines the birds. And given their confinement, emus without a travel option become subject to a host of novel problems, not the least of which is predation. In the bush, they have options as to where to breed, and their continued existence in the bush amid foxes, dingoes, snakes, and goannas shows us that they exercise their options well, at least in arid Australia. But when an emu is locked into a farming situation, she becomes dependent on human intervention. Confinement makes the emus' progeny easy targets for enterprising nest raiders, and this in turn leads to the need for more electric and wire-mesh fencing to keep the eggs and chicks from being devoured. On large emu farms, the chicks are separated from their parents. Eggs are collected daily and hatched in sheds where incubators control for temperature and humidity. The generational connections between male emus, which perpetuate nurture down the lineage, are severed, leaving cohorts of birds separated by age. Chicks will never learn to forage by studying their fathers' example; they only need to learn to use a feed bin. In the wild, their curiosity might be sufficient to find them a diversity of food, but on a farm there is no need. Mathematical models provide the exact amino acid requirements for particular age ranges, so the choice of what to eat is made for them.[42] The thickness of yellow fat beneath an emu's skin is no longer tied to the drought and rain of the outback or the ripeness of fruit in particular places.

The connections between parents are also severed. As with crocodiles, artificial insemination is replacing mated pairs with "positive human-ratite" relationships.[43] The means by which emus choose their mates have lost relevance.[44] Allowing birds to pair off and set up territories within a segment of fenced paddock is too old-fashioned for emu farmers. When birds don't get along, the problem birds are separated out and removed from the farm's gene pool, making for an increasingly tolerant population of birds who can be kept in ever higher densities. Concentrating emus on small plots of land can subject them to diseases and internal and external parasites, but these are separated out with vaccines and worming tablets and by simply removing diseased emus.[45] And the problem of aggression and fighting among emus in confinement is addressed by separating claws from feet. This is not a practice at Emu Bliss, but at some farms the distal phalangeal joint, the middle toe, is removed from chicks so they don't damage each other's skins and the skins of the farmers.[46]

Emu farming is unmaking writ large, binding domestication and late modernity into a unified paradigm. Birds who would otherwise wander are fenced off from their lines of travel, shipped across oceans, contained in paddocks, wormed, vaccinated, and declawed, severed from the bush and dragged impotently kicking into global networks of capital, commercial feed regimes, robotic semen collectors, and globe-trotting viruses.[47] Sheltered from snow in the United States and monsoon rains in India, they are being adapted to fenced lands and feed bins. Their bodies are measured in liters of oil, kilograms of flesh, and square meters of skin.[48] No longer nomads of the bush, farmed emus are denizens of landscapes being unmade. These processes all select for a different kind of bird, one who breeds at a particular time and who adapts well to confinement; who tolerates other emus and ejaculates in the hands of men; who needs no father but depends instead on the protection of fences. The emu of the great unmaking will have a markedly different ecology from the bird of the bush, the bird of Aboriginal dreaming stories, and the bird that our illuminated society is making increasingly difficult to see in the night sky.

6.

Kangaroos

The poor starved sheep, wondering where all the grass had gone, and why it did not come after the rain as of yore, were unable to copy the hardier kangaroo, who, if he never grows fat, can live on country where the domesticated animal must die.
—C. W. NEVILLE-ROLFE, "Some Birds and Beasts"

If anything epitomizes unmaking in terms of plants and animals, it must be the acclimatization societies of the 1800s. These were French and English gentlemen's clubs, funded by the aristocracy and intent on reshaping the raw material of colonization into human mastery over "nature" for the betterment of all. Riding high on a wave of Enlightenment philosophy, acclimatization societies saw animals as things that could be taken from their ecological contexts and moved around the world to be used for Man's benefit. This view of animals reflected the science of the time, which was largely blind to the interconnectedness of the world's systems. Little wonder, then, that acclimatization societies were integral to the establishment and success of zoological gardens—the ark royales of unmaking. But they were also accomplices to massive acts of environmental destruction and, not without irony, to the hardships wrought upon farmers by introduced species. As Michael Osborne says, "Acclimatization is intimately entwined with the rise of modern imperialism and with the marginalization and alteration of indigenous ecosystems and peoples."[1] The Acclimatisation Society of

Victoria was founded in 1861. It was largely funded by the Victorian government and a local newspaper baron named Edward Wilson, who was also a member and an energetic proponent of the society and its aims. In fact, Wilson was behind the introduction of alpacas and llamas to Australia. In October 1861, Wilson's newspaper reported on a dinner held by the Acclimatisation Society of Victoria and attended by various gentlemen of standing. The dinner was a culinary test of Australia's fauna. A celebrated chef served up wallaby, possum, echidna, fowl, and indigenous aquatic species, all with a view to whether these animals had potential in local and world markets, and in the assumption that they could be taken up and farmed at the will of any man.[2] These men had no inkling that the intricate webs that sustain animals as parts of ecologies might undo any attempt to farm them. Their main criterion for the suitability of a species for domestication was whether the meat could be complemented by a béarnaise sauce. It was a given that kangaroo would be served. The haunch of kangaroo at the head of the table suggested that these men saw potential in these abundant game and that, if the meat were given an acceptable name, kangaroo, among other species, could be farmed and sold in the markets of the world. They were halfway prescient. The acclimatization societies are long gone, but kangaroo products, mostly meat and skins, are sold throughout Australia and exported to fifty-five countries worldwide. Directly and indirectly, the kangaroo industry has created four thousand jobs, and in terms of production it is worth AU$175 million to the Australian economy, with a further AU$25 million derived from ancillary benefits. These ancillaries—reduced damage to agricultural land and fewer road accidents—derive from a peculiar fact about the industry: kangaroos are not farmed. Australia's kangaroo industry constitutes the largest land-based hunting industry in the world.[3]

For colonial Australians, kangaroo hunting has served different purposes at different times. At the time of the first settlement, Europeans killed kangaroos for meat and skins. Those first colonists were effectively castaways on a very big island, and their imported animals and crops were still taking root, so killing kangaroos was a matter of survival.[4] But once livestock farming was established, kangaroo hunting became more of a pastime: kangaroos were good sport. When English royalty visited Australia, they joined some of these kangaroo hunts, and had it not been for kangaroos' preference for eating grass, these gamey macropods might have been nurtured like deer on English estates. But Australia was rapidly becoming dependent on sheep and cattle, and kangaroos were seen as competitors. In 1880 they were given vermin status, bounties were offered, and kangaroos, along with countless other hopping marsupials, were killed by the million. Some, such as the eastern hare-wallaby, were hunted to extinction, but kangaroos are resilient, and they persisted. Their killing continued through the early twentieth century as a market opened up for skins, and in the 1950s trade in kangaroo meat developed, which included export for human consumption.[5] At that time, settler Australians were averse to eating kangaroo, but they couldn't legally buy kangaroo meat for consumption even if they'd wanted to.[6] It was legally deemed fit for human consumption in Australia only in the 1980s. During that decade the meat became acceptable as food and a model was developed whereby kangaroos could be killed in large numbers and the meat and skins sold in domestic and overseas markets. Killing kangaroos was also supposed to reduce grazing pressure on sheep and cattle farms. Still, there were conservation concerns, and, not wanting to repeat the mistakes of the past, state and federal governments controlled the kangaroo industry tightly so as not to wipe out the resource it depends upon. To this day, it is a heavily regulated industry. Kangaroo populations are estimated using aerial transect surveys, harvest quotas are based on kangaroo numbers, and population trends are predicted on the basis of expected droughts and rainfall.[7] Shooters must purchase tags, and each kangaroo carcass processed must have a tag attached. Animal welfare is also an issue, and professional shooters must pass a proficiency test to demonstrate that they can competently make a head shot. There is also a lot of concern about the pouch young of killed female kangaroos; they are dispatched with a blow to the head. But this issue has diminished now that meat processors have a male-only policy. Apparently, the skinning process causes the bacteria in a female's pouch to contaminate the carcass, so the vast majority of kangaroos killed for meat and skins are now males.[8]

I'm interested in kangaroos precisely because they are not currently farmed. Superficially, this is due to the low value of kangaroo meat. Landholders, government agencies, and kangaroo meat companies will all tell you that farming kangaroos is not feasible because the financial returns do not justify the outlays. But I think this is only part of the picture. Certainly, the idea of kangaroo farming is as old as colonization, and in the 1960s there were some serious attempts to make it viable. The Commonwealth Scientific and Industrial Research Organisation (CSIRO), for its part, maintained a kangaroo farm for research purposes. On a ten-acre property, the CSIRO kept a colony of six species of kangaroo and eight species of wallaby.[9] The

CSIRO established pasture for the animals, dressed the grass with superphosphates, carried out yard rotations, and supplemented the roos' diet with hay, oats, and mineral supplements. Actinomycosis, a bacterial disease known as lumpy jaw, was a problem, and the researchers noted that studies were needed for purposes of learning how to raise these marsupials successfully in captivity. They even experimented with fostering young kangaroos with nonbiological mothers. The only element missing from the research was a concern with costs and returns. The cost of high fencing, feed supplements, labor, and medical care would have outweighed any returns from the meat, were it to be sold.

It fell to the National Parks and Wildlife Service (NPWS) to assess the economic feasibility of farming kangaroos. The NPWS had received a lot of inquiries about the possibility of farming kangaroos and decided to investigate its viability. Reproduction and growth rate were important considerations. Interestingly, the benchmark against which the roos were measured was sheep, compared to whom kangaroos are, unsurprisingly, less productive, although a given area of land can carry twice as many kangaroos as it can sheep. Ironically, the NPWS found, the optimal kangaroo farm will have only one male per group of females, so as to prevent fighting. But unless the pouch bacteria that compel a male-only harvesting policy can be overcome, the majority-female farming model might present a problem. Another problem was the movement of kangaroos across property boundaries. While there were few data on the subject, the assumption at the time was that regular fencing would be inadequate, so that anyone wanting to keep their kangaroo stock from drifting off would need to build fencing 1.8 meters high. At AU$2,200 per kilometer of fence in 1974, this meant an exorbitant cost. As far as diseases went, researchers had little to go on. Capture, handling, and overcrowding were clearly factors in causing and spreading disease, another strike against farming kangaroos. The market for kangaroo products presented additional problems. At the time, it was fluctuating, and the export market was especially precarious because of concerns in other countries about contaminated meat. Markets were drying up in some places as easily as they were opening in others. The Parks Service also identified problems with ownership. Because the Crown owns Australia's wildlife, farming kangaroos raises issues of responsibility and regulation. There is little incentive to farm animals that one cannot own. The study concluded in 1983 that "kangaroo farming, in either intensive or rangeland guise, is not a feasible proposition at this time."[10] That statement holds true to this day.

It might come as a surprise, then, that kangaroos have been farmed successfully in Australia for a very long time. When surveyor Major Thomas Livingstone Mitchell struck a path to the Gulf of Carpentaria in the 1830s, he contemplated the landscape as an interdependent ecology. He saw Aboriginal people, kangaroos, grass, and fire as an ecological system in which each element depended on the others. If one component were to be removed, he suggested, then the landscape would be forested as thickly as New Zealand or America.[11] What Mitchell was describing was the intensive land management practiced by Aboriginal people, referred to as fire-stick farming or fire-stick ranching, in which kangaroos were the primary product of the farming and ranching practices. While not the only way to manage kangaroos—according to some accounts, Aboriginal people also herded them into battues or cut the Achilles tendons of captive ones to stop them running off—the fire-stick method was the most prevalent.[12] The basis of fire-stick farming is kangaroos' predilection for young grass shoots, the very things that first appear after an area of land is burned. Australia's Aboriginal people knew this, and they deliberately burned the land in order to provide graze for kangaroos, whom they then speared. The burning was carefully timed and was ordered by tradition and clan associations with land; it was a very deliberate and managed method of controlling the movements of kangaroos across the landscape. Not only did it attract kangaroos to grassland, but burning hill country also compelled kangaroos to move toward the plains, where they were easier to kill.[13] The evidence of kangaroo nets and battues tells us that these people could have used fencing, but why go to the trouble? It was much easier to set fire to a particular area of land at a predetermined time and come back later to harvest the kangaroos who were drawn to the sprouting grass. This has been called mosaic burning—the creation of a landscape that holds different ecotones (transitional regions between biological communities) at different stages of regeneration. Not only does fire-stick ranching control the movements of kangaroos; it draws wildlife away from planted foods and makes for ease of movement across the landscape. It's also been shown to create attractive landscapes for other game, such as monitor lizards and emus, and to encourage certain fruiting plants.[14] Bill Gammage's book on the subject presents many accounts from the early European colonizers who encountered these managed landscapes, all of which note the resemblance of the Australian bush to a European "gentleman's park."[15] Over millennia, fire-stick farming had created rich and fertile soil covered with lush pasture and dotted with trees and hopping marsupials.

For precontact Aboriginal people, kangaroos were a staple. The stories collected by Ronald and Catherine Berndt give us some idea of how important they were—or at least how much people desired them. In story after story we find dreaming beings variously hunting, trapping, skinning, cooking, and eating kangaroos, disguising family members as kangaroo meat, fighting over kangaroos, and stealing the carcasses from one another. What's remarkably consistent among these stories is that kangaroos are seldom protagonists; in fact, in only one of them is a kangaroo the main character. In this Gunwinggu story from Arnhem Land, Kangaroo appears in a dreaming in his original humanoid form.[16] He asks Dog to paint his body and Dog does so, making him look like a kangaroo. Dog then asks Kangaroo to return the favor and paint him. Kangaroo complies but makes a balls-up of the job, which leaves Dog looking butt-ugly. Dog becomes angry and chases Kangaroo across the landscape, creating various physical features as he goes and bequeathing to his descendants an uncontrollable urge to chase and bite kangaroos. As for the other dreaming stories in the Berndts' collection, however, kangaroos are typically functionaries whose role is to move the stories along. They are shared targets, currency, and meat; they give the protagonists not just something to eat but something to hunger for, and something over which to come into conflict. In a society without money, kangaroos were the objects of greed, betrayal, and deceit, as well as a means of finding common ground and reconciliation through the sharing of their flesh. They serve as functionaries because Aboriginal people really liked—and still like—to eat kangaroo.[17] In terms of diet, Jon Altman gives us some idea of how important they were. He spent a year faithfully recording the species of plants and animals used by the Gunwinggu of the Mann and Liverpool Rivers region in the Northern Territory.[18] While his data don't paint an accurate picture of precontact diet—the people used motor vehicles and guns and also ate store-bought food—they give us an idea of the importance of kangaroos. The movements of the Gunwinggu bands and the foods they ate were seasonal. In late March, they moved into the tidal reaches of the Liverpool River to catch fish, a month later they were on the Tomkinson floodplains harvesting birdlife, at the end of the dry season they were harvesting sugarbag honey, and during the middle of the dry season and early to mid-wet season they harvested mammals. Unfortunately, Altman doesn't provide a breakdown of exactly which species of mammals they harvested. Along with various macropod species, there were feral cattle, buffalo, and pigs. Seasonally, however, "mammals" constituted between 50 and 87 percent of the total bush foods consumed by these people, and in precontact times the majority of mammals were kangaroos and wallabies. Altman also notes that they hunted kangaroos with fire drives, effectively combining burning with harvesting. But perhaps the use of guns by these people in modern times exaggerates the significance of kangaroos in precontact diets. Anthropologist Michael Pickering shows how hunting kangaroos effectively is in no way dependent on guns. His example of "salvage ethnography" among the western Garawa people documents accounts of traditional hunting and food-preparation methods that are no longer used.[19] Hunters traditionally smeared their bodies with ochre to disguise themselves and mask their smell. Like Bininj hunters of emus, they held branches in front of themselves to get to within spearing range of kangaroos. They targeted the lower body of the animal, and once speared, a kangaroo was either dispatched and eaten immediately or, if only incapacitated, left to be consumed at a later time.[20] Another technique, this one for catching hill kangaroos, was to find a position above the spot where the quarry was resting. A hunter took a stone, wiped it under his arm, and threw it upwind of the kangaroo's resting place. The sound and smell from that direction caused the kangaroo to flee toward the hunter, who was ready with his spear. The Garawa also used fire to drive kangaroos toward groups of hunters, and apparently dogs and hunters joined forces to drive kangaroos onto boggy ground, where they were speared. Pickering's informants told him that one macropod per hunter per week was the minimum catch, while a good hunter could catch one or two animals per day. On a per capita basis, that is a lot of meat. And unlike every other species consumed by these people, macropods were hunted all year round and at a consistently high frequency. Fish were only caught from July to November, sugarbag only collected in large amounts from August to November—but macropods were a year-round constant.

Still, kangaroos could be much more than meat, and some Aboriginal people have very close associations with kangaroos. This is highlighted by a particularly eye-watering mode of cultural body modification: subincision. This practice requires that a man cut a slit into the ventral surface of his penis at the urethral orifice, and over time reopen and extend the slit as far as the base of the penis so that the entire urethra is made into an open channel. While this form of genital modification is not unique to Australia—it is sometimes done in other countries for therapeutic purposes—the reasons why Aboriginal men do it emerge only within the context of the Australian landscape. Among many language groups of the Western Desert and Central Australia, subincision is, or was, obligatory for all males. This practice attracted the attention of not a few social scientists, who speculated on the reasons for its widespread persistence. In the 1960s, psychiatrist John Cawte and his colleagues set out to get to the root of subincision practices among the Walbiri of Central Australia. Initially, they focused on which aspects of subincision—known locally as *burra*—seemed practical to Western eyes. Based on a study of the literature and a bit of "armchair ingenuity," they arrived with a bunch of hypotheses about various benefits that *burra* might confer, and some spiritual reasons for why the practice persists.[21] These included speculation that it might be more hygienic than a regular penis and that it might confer increased sexual pleasure for men and/or women. They even applied a Freudian approach and suggested that it might express vulva envy, and/or that the constant reopening of the incision might be seen to mimic menstruation. But they realized that their frames of reference were too narrow when they were told stories of how the *burra* practice had its genesis in the dreaming among the beings from whom the people were descended. One of the beings who feature prominently in *burra* stories is Kangaroo. In one story, Kangaroo meets Marsupial Mouse, and while the two are camping together, Kangaroo shows Mouse something the little marsupial has never seen before: he draws blood from his *burra* and drips it on Mouse's shoulders. That's one way to liven up an evening by the campfire. Kangaroo instructs Mouse in how to make his own *burra* and teaches him the songs that go with it. In another story, two Walbiri kangaroos discover a flint knife wrapped in paperbark and use it to create a *burra*. In a third story, two kangaroos find a boy being prepared for an initiation ceremony. They blow flint knives out of their noses and show him how to cut a *burra* with one. After the operation, they tell the boy, "You're free like us."[22] The story ends with some crucial information from the narrator: the *burra* is copied from the kangaroo, who is father of the peoples of Central Australia. This last bit of information sent the researchers back to their biology texts, in particular the sections on marsupials, where they learned that indeed marsupials often have a bifurcated glans, giving the penis a split appearance. For their part, kangaroos exhibit the urethral opening not at the tip but along the ventral surface of the penis, and they also have a groove running along the length. So while the therapeutic, social, and practical aspects of *burra* are important to the Walbiri, the stories speak of something broader. When a man slits the base of his penis and opens up his urethra, he is, in a most emphatic, literal way, inscribing kinship with kangaroos onto his body and honoring his human/kangaroo ancestors who created the land. The kangaroo people of Central Australia don't subscribe to totemism in its abstract sense; rather, they painfully reconfigure their bodies to manifest kinship with the animals who bear their relatedness. It's a far cry from the colonial view of kangaroos as vermin.

I'm heading west, over the mountains and away from the humidity of the mid-north coast, into the flat scrubby expanses of the New South Wales outback. My destination is the town of Bourke on the Darling River. A former boomtown for the wool industry, Bourke's fortunes have waned as the pasture that nurtured multitudes of sheep has steadily degraded, and the depredations of wild dogs, scrub, feral animals, and kangaroos have had an adverse impact on the enterprise and optimism that characterized colonization. When Europeans first explored this area, they encountered fertile lands and an inconvenient Indigenous civilization. Thomas Mitchell described the countless campfires seen through the trees at night, the women who fished the rivers, the roads leading in all directions. He described a plain along the Darling River where "ricks or haycocks extended for miles," where the local people had harvested native millet for threshing.[23] The remains of the fish traps at Brewarrina convey the importance of this river in more than economic terms. The traps were a shared resource; according to folklore, these linear arrangements of rocks were created by the Ngyamba with the help of an ancestral hero, Baiami, and his giant dingo companions.[24] But they were used and maintained by four groups, the Ngyamba, Wangkumara, Wayiliwan, and Kamilaroi. Evidence of changes in river flow suggests that the fish traps may represent thousands of years of communal use of a shared resource.[25] Of course, the fish traps meant little to the colonists who stole the rocks to build a crossing upstream and then cleared others to allow watercraft to pass. But like the Aboriginal people who maintain them, the traps persist in spite of colonial efforts to ignore their statement of ownership.

The early colonists were at least justified in their optimism about this part of the country. Intensive management by the Aboriginal people had resulted in a rich and fertile country that attracted abundant game. When Charles Sturt arrived in the region in the 1850s, he described an expansive plain of grass "up to the horse's middle" as he rode through. And Captain William Randell made comparisons with the rich pasture of the mother country, describing grass that could only be walked through with difficulty, "as thick and close in the bottom as the meadows of England."[26] Indeed, this is the region where Thomas Mitchell made his prescient observation about the interdependence of people, kangaroos, grass, and fire. As it transpired, colonists set about removal of three of those components, because all they wanted was the grass. The result was, and is, a massive ecological (and social) breakdown, which is ongoing. The Aboriginal people were either murdered in their camps, herded into a reserve at Brewarrina, or put to work on the newly established lots of privately owned land. The kangaroos were inimical to wool production for they were seen as competitors of sheep; the 1880 Pastures and Stock Protection Act declared them pests in need of extermination, opening the gates to widespread killing. Fire was considered a threat to livelihoods, and in its absence the grass inevitably dwindled. An addiction to sheep obscured the degradation of the land in the colonists' eyes and wrote a noncyclical story of boom and bust. Without periodic fires, the scrub known locally as woody weed outcompeted grass, and the pastures became dustbowls.[27] Bore water troughs increased the carrying capacity of the land exponentially, enabling rapacious woolly destroyers to clear vast swathes, leaving nothing to blow in the wind but dust, which swept across the land in great clouds. By the 1990s, the land could sustain only half the former number of sheep. Farming regimes, poisoning, shooting, introduced animals, and the lack of burning had caused the extinction of twenty-seven of the eighty-eight mammal species of the region.[28] But despite the upheavals, and to some degree because of them, kangaroos thrived.

If kangaroos were flies, the road to Bourke in outback New South Wales would be one long strip of asphalt fly paper. I lost count at a hundred, but I must have driven past three thousand kangaroo carcasses in varying states of decay—such is the sheer quantity of kangaroo roadkill. And that is just one road among many in the Australian outback. The confusion that kangaroos experience in the face of bright lights makes them hopeless pedestrians, and their propensity for jumping in front of speeding vehicles has profound effects on people's lives in these parts. Traveling these roads between dusk and dawn makes for a high likelihood of hitting or being hit by a kangaroo, so locals either avoid driving after dark or attach roo bars to the fronts of their cars and trucks. But roo bars are not always enough protection, as some kangaroos collide with the sides of cars. A local man complained to me that two brothers who borrowed his car brought it back with a massive dent in the front quarter panel where a kangaroo hit them side-on. They in turn blamed him for demanding the car back in a hurry and forcing them to drive back to Bourke at night. Insurance companies in particular see a problem with the high number of incidents, so we may add them to the list of entities that see kangaroos as pests that must be controlled. Another feature of kangaroos adds to the tragedy of driving outback roads: females almost always have a joey in the pouch. I had firsthand experience of this one night as I was driving north. Just after sunset, I watched a kangaroo hop across the road ahead of me and into the path of an oncoming road train.[29] I silently shouted at her to jump clear, which she did—into the path of my car. I jumped on the brakes but still hit her hard. As she bounced off the front of my car, a joey spilled from the pouch. The female kicked herself down an embankment with one good leg while the joey died on the road in front of me. It's an experience that's difficult to forget.

Roadkill is fresh in my mind as I knock on the door of a Bourke guesthouse. By Bourke's standards it is large—formerly a bank built in the 1800s—an imposing structure that would have left no doubt that the colonists were here to stay and intended to make a lot of money from the land. The current occupants have softened the massive stone building with lots of wall hangings and potted plants, a shady garden of veggies, intricate metal sculptures, and chickens running around the yard. I go to reception, a massive space—formerly the bank's customer service area—occupied by a lonely desk in a dark corner. The guesthouse owner, Christie, appears and beckons me outside to see the long-term residents. When not occupied with running the guesthouse, Christie and her partner care for young kangaroos. She takes me to an adjoining lot, and as soon as we enter through the gate we're surrounded by joeys. It's feeding time, and Christie has an armful of bottles for them. She sits and the joeys crowd around hungrily. There's a bunch of reds, a couple of eastern greys, and one western grey. They share one characteristic: they're all orphans. Their mothers were all killed on the road, and whether conscientious truck drivers or concerned tourists, the people who found them—in most cases the people who killed their mothers—brought them to the local wildlife rescue or the local vet, who passed the orphans on to be rehabilitated. She takes me to a second paddock where a line of cloth sacks hang from a frame. These contain joeys younger than seven months who are still pouch-dependent. Christie's partner, Chris, explains to me how easy it is to manage the kangaroos. "They're not aggressive at all; they don't try to get out or bite or scratch. Really, they're just beautiful, placid, intelligent creatures." The couple raise the joeys on milk in a small yard and eventually release them at Gundabooka National Park, where a park officer friend feeds them and takes the edge off their transition to the bush. Chris and Christie choose not to name their joeys because a great many come into their care and they inevitably release them. They say it's unhealthy to form too much of an emotional attachment. I ask what they think about the idea of farming kangaroos. "I think they should be farmed," says Chris. "There's so many of them that it makes perfect sense. But the fencing is so expensive." What about site fidelity? Mightn't they just stick around because they're attached to a particular place? "Normally, yes," he agrees, "but in this country they'd follow the rains. If it rains over there on the horizon, they're going to follow it to where there's pasture." It's an interesting place, this kangaroo orphanage. On the one hand, it's a clear expression of the human desire to nurture, and Christie is a self-confessed nurturer. But, on the other, it shows how these kangaroos quite readily lend themselves to the domestic sphere. These joeys are very comfortable with their human caregivers and not particularly disturbed by me or my camera. In this sprawling backyard of Bourke, I see a very domestic bunch of kangaroos.

In another backyard—this one at the Port of Bourke Hotel—I meet with Fiona Garland, a field officer with the Local Land Services (LLS), a state government authority dedicated to reconciling farming with Australian ecologies. Fiona is passionate about kangaroos and seeks a place for them in the social ecology of Australia. In 2016, she convened a workshop on kangaroo management out of which emerged a task force comprising representatives of the LLS, the Royal Society for the Prevention of Cruelty to Animals (RSPCA), the Australian Veterinary Association, and a host of other groups. She outlines some of the major issues involved in kangaroo management. "The main issue is the price of kangaroos. Obviously it's a little more complex, but at base it's an issue of price. If kangaroos are not commanding a high price, then there is little incentive for shooters to come out and harvest them, and so no incentive at all for landholders to intensively farm them. Goats, on the other hand, are profitable. Ten years ago they were a serious problem. Then an export market opened up and they became a boon. Landholders were happy to have them on their land. Considering they were fetching $150 per head, it was feasible to build trap yards. Harvesting began and goats became a nonproblem."[30] Kangaroos are also considered a pest, but not one that is easily managed. "Kangaroos are owned by the Crown, but the Crown doesn't take responsibility for them," Fiona tells me. "It falls upon the landholders to control kangaroos, but as they belong to the Crown, landholders can't manage them as a pest species. They're protected. So a landholder has to get permission to harvest or cull them." For some, this is an onerous task, as the process of getting a permit to "shoot and let lie" (a permit to shoot kangaroos for noneconomic purposes) requires a trip into town. "Some of these landholders have to travel three hundred kilometers to town to get a permit." This will all change when the permit system goes online, but it underlines the difficulties that farmers face when they have a kangaroo problem. Another issue connected to market value is social license. "If Australians and people overseas continue to see kangaroos as iconic, cute, beautiful animals," says Fiona, "then it's always going to be hard to create a strong market for kangaroo products." But even if that hurdle is crossed, there is still the issue of animal rights groups that oppose killing kangaroos under any circumstances. Also, Aboriginal groups are unhappy that they've been excluded from the discussion. Fiona's view is that "we need to create the social capital to sustain a kangaroo industry." So what does she see as the best way to reconcile kangaroos with sheep and cattle on the rangelands? "Some people suggest that the rangelands be left alone—that they be turned into national parks," she says. "In fact, some people think they already are. But they form a major percentage of productive land in this country, and the demand for products of the rangelands is going to increase, not decrease. That's a reality we can't just turn our backs on." In terms of food security, the rangelands will be crucial, but with only a few hundred millimeters of rain per year, people won't be growing tomatoes out here anytime soon. "We need to promote the positives," says Fiona, the fact "that these are free-ranging animals with little in the way of interventions such as drenching and so on. They have environmental cred. As far as managing wildlife and livestock, the answer is fencing. Control the movements of these animals by fencing and by control of water points. Create systems for commercial and noncommercial culling and work towards biodiversity and healthy landscapes." I ask whether the LLS has considered burning. "Yes, we're proposing cultural burning, but this is a problem in the rangelands. Not only are a lot of landholders quite conservative in that they see fire as a bad thing for the land; they're also involved in carbon-sequestering projects which won't let them burn." I'm getting an idea of just how far Australia may be from farming kangaroos.

Australians have an expression for anyplace that's way out in the outback, in the middle of nowhere: "back o' Bourke." I'm literally there—at a place called Fords Bridge, some seventy kilometers northwest of Bourke. Fiona suggested that I go to a meeting here of Landcare Australia, a conservation nonprofit, so I've driven through the red dust and scrub. The town is modest.

There's a pub on one side of the road and a shed on the other. The pub has an air conditioner and, more important, beer, so it's a focal point for the local populace. We sit around a table in the pool room while Alice, the Landcare officer, discusses programs and funding avenues for farmers who want to improve the ecological integrity of their land. I get the impression that the farmers have the government's ear, but I also gather that they're struggling. Kangaroos and woody weed (native scrub) are constantly encroaching on their land, and much of the discussion is about mitigation. It's a very convivial meeting and everyone seems dedicated to working together and with the government to iron out the problems.

After the meeting breaks up, I sit down with a couple of farmers to talk about kangaroos. Steve and Helen both own large acreages northwest of Bourke. Much of their land is overrun with woody weed, which although it is native they see as invasive because it's a relatively recent phenomenon. Helen's mother was raised in the area, and she remembers only small outcrops. "They used to laugh at the neighbor, Mr. Hoskins, who went with a bullock team and put a chain around it. They couldn't see why he'd bother with a scattering of shrubs. Now farmers need pairs of bulldozers with chains run between them to clear the stuff." And the scrub attracts kangaroos. While it doesn't provide pasture—the farmers do that—it does give them cover. Kangaroo shooters aren't interested in operating in areas where there's a lot of scrub. Steve tells me that shooters won't visit his property for this reason. "They don't want to come out to somewhere, go *bang*, 'Oh, where's that one?' and get three punctures gettin' to it, get bogged, and pick up one roo. They want to go to an open area, go *bang, bang, bang, bang, bang*, and go pick up ten roos." It's just too difficult to shoot and recover kangaroos in thick, scrubby terrain. As a result, Steve and Helen have hundreds of kangaroos on their properties, which they see as competing with sheep for pasture. Both farmers express an appreciation for the beautiful, iconic animals; Helen even has a pet roo named Skippy. But both are pragmatic about farming practices and see kangaroo control as a necessary evil. Because kangaroo shooters won't work their properties, they have to buy shoot-and-let-lie tags from the government and do the shooting themselves. But in these cases the kangaroos can't be processed for meat; instead, they are shot and left to rot where they die, as Helen says, "wasting all that beautiful meat." The reason that they can't be processed is the high cost of specialized equipment that meets government hygiene standards and the bureaucratic hoops farmers have to jump through in order to harvest kangaroos commercially. Not to mention the ungodly hours involved in shooting kangaroos, who only emerge from cover after dusk. All of this might be worth the effort if the returns were decent—farmers recognize they need to diversify in hard times—but at sixty cents per kilogram for a gutted, processed kangaroo, nobody's going to pour themselves and their land into such a venture. "After a contamination scare, the Russians stopped buying kangaroo meat and just about killed the industry. Western Landcare are trying to get in with politicians and get things happening—it's obvious we have a problem—but we need to market them." Steve says emphatically, "If there was a market for it, there'd be no roo problem. If it was in the supermarkets and available at half the price of lamb, I'm sure it would sell. Look at the goats."

I had looked at the goats. I'd seen scores on the drive here, darting into the scrub at the sound of my car, scampering across the road ahead of me, poking their heads up from behind bushes. Anyone passing through would think that feral goats are a problem here, but in fact they're far from it. Whereas kangaroos fetch a paltry sixty cents per kilo of dead weight, goats have been earning farmers upward of three dollars per kilo for a live goat, and all a farmer has to do is round them up and take them to town. "It's relatively easy money," Steve tells me. "I can go out with a motorbike and a dog, have a rough yard set out in the scrub, and get a mob of goats. I don't have to do anything except put 'em on the trailer, drive 'em to the depot, and drop 'em off. A little bit of petrol to drive them to town and that's it." And goats are much easier to muster than kangaroos. I get a firsthand perspective on this as I drive through open country where goats and kangaroos graze the same area. When I get too close to kangaroos, they bound off, but not in any sort of cohesive way. It's every roo for himself. The goats, by contrast, are solid. When spooked, they mob up and move off as a herd like a single entity. But price is also a key factor. Once upon a time, feral goats were almost worthless, fetching only five dollars per animal. They were seen as a pest competing with sheep for valuable pasture. But as the price rose, goats became a valuable resource. "Goats were vermin," Steve tells me. "People used to come out from Melbourne to shoot 'em. They might have shot a hundred. Now, if someone shot one goat, you'd shoot *them*." Helen concurs. "They're such a large part of our income now," she says. "They're putting food on the table and paying the bills. We need our goats. They paid for the air-conditioning in my house." So lucrative is the goat market that even the old-school farmers are getting into goats. "All the blue bloods—the people born and bred with the family generational thing in this region—would have never touched a smelly old goat. Well, now they're like the rest of us. The price is there; they're worth farming." But goats are a different beast from sheep in terms of farming. They're not considered stock in the same sense as sheep. Goats can come and go—they cross property boundaries, and as individuals they belong to whoever owns the land they're on. When Helen has to list her assets on any sort of paperwork, she can only estimate the goat numbers. It's a curious sort of arrangement, because goats have value as stock but ownership is fluid. "You never know how many you've got. They might be there today and not tomorrow because they've jumped over that fence and they belong to somebody else. I used to raise the orphan ones, give 'em ear tags and let them go. But you can't control them." At the moment, farmers aren't going to the expense of putting up the fencing required to keep goats in. Yet things may well change. "It's got that way now where the price has gone up so high that they're trying to control where the goats go," says Helen, "but it's expensive for the fencing. That's the only reason why they haven't done all of their fences." It will be interesting to see how the emerging goat industry pans out as reliance on goats increases. But one thing is certain: as long as there is a market and goats are easier to manage, the prospect of farming kangaroos fades into the heat haze of the plains. "We've all contemplated it at some stage," Helen tells me. "'Gee, I wonder if it's worth doing the roos.' No, it's not. You get one flat tire, there's your profit off the back of the pickup gone."

In talking with local farmers, I'm beginning to see how central kangaroo shooters are to the kangaroo industry. They are the ones earning an income from kangaroos—the farmers only benefit from the kangaroo industry by having grazing pressure reduced. These shooters connect landholders, processors, distributors, animal welfare groups, and government organizations, and all of these are connected to kangaroos via the shooters. The kangaroo industry depends on the shooters, but they are also the ones who shoulder much of the burden in terms of kangaroo welfare. And the shooters have the final say about where they are prepared to work, so they are the ones who decide where kangaroos will be harvested. After attempting to speak with a shooter through various channels and coming up empty-handed, I'm ready to give up on finding one who is willing to talk to me. But then I manage to get the number of a shooter named Ross Kemp from a contact in the LLS. I call the number and a woman answers the phone. She says Ross is still asleep. It's three in the afternoon. I apologize and ask when to call back. "Give him another hour and he should be up and ready to talk to you." I decide to give him two hours, but he calls back an hour later, ready and willing to talk.

Ross is a very open, amenable guy. He tells me about the long hours involved in shooting kangaroos. "On a normal night I go out at around 7:00 P.M. and work through to 7:00 A.M. I go home, get cleaned up, have breakfast, and go to bed. I get up in the afternoon and get onto prepping and cleaning for the next night." Sounds demanding. "I'm a motor mechanic by trade. Sometimes, when it gets a bit much, I take a break and do some mechanical work for a bit." I ask him how many kangaroos he typically shoots in those twelve hours. "The property I'm working at the moment, I reckon there'd be four thousand in the one paddock," he says. "I've shot over a thousand this month alone at about sixty per night. They have places where they congregate—usually where the rain band has gone through—and I go to those places." Do the landholders contact you, or do you approach the ones where you predict there'll be roos? I ask. "Usually they approach me. They tell me where the roos are and I can see the numbers. A lot of the time they overestimate, probably because they feel like they're under siege." I ask if there are landholders Ross won't work for. "Yeah, there's some difficult people, but that's life. Some graziers jump up and down asking me to come out, and I get there and there's hardly any roos. Or they might have high expectations that I just can't fulfill." Do they contact you ahead of destocking paddocks? "Oh, yes. When they're planning to destock they'll give you a call and ask you to come out. It's a pretty good bet when someone destocks that they'll get a lot of roos." I ask him about the factors affecting kangaroo populations. "On the one hand," he tells me, "there's the exclusion fencing. You really notice it after a while because they can't move to where there's pasture. Inside the fencing the roos steadily lose condition, and within twelve to eighteen months they vanish. Still, they're adapting and learning to dig under the fences. One bloke spent a fortune putting up a fence on a clay pan, and the roos ended up digging their way under where there was a post. The post fell over and the farmer was dirty about that. This is the one way that goat management is affecting kangaroos: you see roos dying in paddocks that are fenced to keep goats in because the roos can't move to where there's feed. The other thing is the buck-only rule. It's seen an explosion of animals. It's really propagated their numbers." Aside from numbers, I want to know, has he noticed any other changes—say, in flight distance? "Definitely, they're making it harder. Dad used to take us out shooting when we were kids. We'd take a low-voltage light and a .22 or .303. Nowadays, you need enough to light up a stadium and you need a .323." As I listen to Ross, I can't help liking him. He's very low-key with his opinions, and considering the long hours he's been working, he's remarkably sociable. So I get personal. It sounds like you're working long hours doing an awful lot of killing, I offer. Is there an emotional toll to all that? Ross is unabashed. "Absolutely. You can't be in this job without it taking an emotional toll. I've raised dozens of joeys here at home because I couldn't bring myself to knock 'em off. But the way I see it is: if I don't do it, then someone else will, and not as properly or humanely. Some graziers poison water sources and the roos die horrific deaths. And there's the so-called sporting shooters who come in and shoot them anywhere, leaving roos to die. I've seen starving roos with broken legs and all sorts of suffering because of sporting shooters." I've done a double take. You raise joeys? "Sometimes it's just too hard. If there's one that's a little bit feisty, I'll say, 'Come on, you' and bring him home. It happens a lot."

My connection to Ross transcends a phone conversation. I eat kangaroos. I'm not particularly fond of the taste—they're quite gamey—but I eat them because I see it as an ethically better way of engaging with my food. I abhor the industrialization of animal production—it's unhealthy in so many ways. In line with the ecologist-philosopher Paul Shepard, I see the depersonalization of animals that has resulted from industrialized factory farms as psychologically unhealthy. The costs to the animals themselves don't even need mention—they are obvious. I don't see veganism as an answer; in fact, I see it as a form of denial. It severs the life-sustaining connections that our ancestors maintained with animals and fosters the same kind of dualistic humanism behind modern animal production and consumption. Veganism and modern animal production are two sides of the same coin.[31] I subscribe to the Bininj view that necessitates an engagement with animals, not as their masters, protectors, or superiors but as their kin. After talking to Ross, I feel like a burden has been lifted. I've been eating kangaroo as diced bits of meat in sealed plastic packages far removed from the furry hopping creatures of the arid Australian rangelands. It's created a niggling distortion that has undermined my ethical perspective on eating kangaroo meat. I'm not engaged with the kangaroos whom I'm eating, or with the people who are killing them. In terms of kangaroos, I'm a consumer at the end of a supply chain. Cashiers, purchasing officers, warehouse workers, clerks, drivers, marketing people, meatpackers, processors, shooters—all stand in anonymity between me and the anonymous kangaroo whose flesh I eat. One of these is an Aboriginal man named Ryan who works as a cashier at my local supermarket. He shakes his head when he scans the package and tells me I'm wasting my money. "Up where I'm from [South East Queensland], we get 'em for free." I'm deeply ashamed, not because of what I'm paying but of where and how I'm getting it. Having spoken to a shooter, I feel I've leapfrogged over all of those anonymous people and connected with the source. In my mind's eye, I'm standing there when that kangaroo is transfixed in the bright light that is the last thing he will see. But I also take heart in my belief that killing kangaroos at an industrial scale for more than forty years cannot expel the humanity from a man. I know now that riding in the truck with the carcasses from a night of killing is an orphaned joey who holds considerable influence over a man with a .323 rifle.

The mountains that divide the rangelands of New South Wales from the narrow strip of land along the coast create a striking contrast. While the west is dry, flat, and sparsely populated with humans, the east is humid, green, undulating, and crowded, at least in Australian terms. Up and down the coast, what were once small holiday hamlets are being transformed into suburbs. Vacationer houses that thirty years ago were affordable even to the working classes have n-doopled in value and become permanent residences for professionals wanting ocean frontage. People from the cities are migrating to the coastal towns, taking advantage of the workplace flexibility that the internet has created. One of these towns is Woolgoolga. Located just north of the halfway mark between Sydney and Brisbane, "Woopi," as it's known to the locals, is part of an urban spread that will soon link Corindi Beach in the north to Sawtell Beach in the south, creating a sixty-four-kilometer strip of coastal suburban development. In Woopi there are cafés, boutiques, and a branch of the national supermarket chain, and within twenty minutes of the town there are hospitals, a regional airport, and a university. There is still a country feel, because the mountains are so close to the coast here that you can always see either forest or ocean. But development is rife. Real estate agents are multiplying and rural land is constantly being opened up for subdivision. Interestingly, a major selling point for real estate in these parts is kangaroos. Along a twenty-five-kilometer coastal strip within which Woolgoolga is central, kangaroos are abundant, and real estate agents often include pictures of the resident roos in their listings. They offer beachside living with services and close encounters with iconic creatures of the Australian bush.

What attracts the kangaroos to this area is grass, and lots of it. The rainfall here is five times that of the rangelands, and the mild climate makes for constantly green, constantly growing grass. It covers lawns, median strips, verges, sports grounds, golf courses, and what's left of the pastureland. It keeps countless professional mowers in business and demands cutting every second weekend. I know this from firsthand experience because I live in this coastal strip and spend two hours every other weekend pushing a lawnmower.[32] It is odd stuff, grass, at least when it covers so much otherwise productive land. We devote so much time and precious oil to keeping it at a respectable height, in the full knowledge that it will keep on growing.[33] It marks yet another mode of unmaking, where we have turned something so potentially productive into something that consumes us. Where grass once sustained the animals that sustained us, it has been reduced to a burdensome decoration. Still, the kangaroos on this stretch of coastline appreciate it, and the people who mow the grass seem to appreciate the kangaroos. The local golf club is a focal point for roos. I ask the president of the club if they are a pest in any sense and he answers, "Not in the slightest. We just have them here and they don't cause any problems. In fact they attract visitors and tourists to the area." The golf club doesn't use them as a selling point, but anyone playing a round of golf can be almost guaranteed to have a mob of kangaroos in attendance. Certainly, there have been incidents in the suburbs involving kangaroos—one where a boy was scratched up by a male kangaroo who thought he was a potential sparring partner—and kangaroos are always being hit by cars, which causes headaches for insurance companies, but by and large the kangaroos are appreciated and there are no public demands to cull the numbers.

Making the rounds of Woolgoolga in the early morning, I see kangaroos everywhere. They congregate at the local high school where the kids throw them scraps of unwanted lunch—probably the healthier items; they assort themselves across people's lawns, amid garden gnomes, where they're eyed suspiciously by little dogs who know better than to try to chase them off; they bound down suburban streets with a nonchalance that hunted kangaroos of the rangelands could never imagine; they scatter themselves across the local soccer pitch—if you execute a sliding tackle on a Woolgoolga midfielder, you can expect poo on your shorts; and they congregate at a place called Look at Me Now Headland just south of Woolgoolga, where there is grass and coastal heather on offer. This place is kangaroo central. Here you will find them sharing the road with people marching up to the reserve on their morning walk, and they are all over the headland, resting, eating, mating, and sparring. The kangaroos are so abundant that the National Parks and Wildlife Service has an information board at the headland car park with a list of dos and don'ts around kangaroos:

> Keep your distance especially when kangaroos are fighting, courting, or with young
> Supervise children at all times
> Watch where you walk so you don't startle kangaroos where they are resting
> Keep dogs out of the reserve
> Never walk towards a kangaroo
> Do not stand up tall in front of a kangaroo
> Do not feed or interact with kangaroos

I leave my car at the car park on the headland and follow the paved walkway that loops around to the south. The morning air is cool, the grass wet with dew. The kangaroos keep the vegetation low around here, making for uninterrupted views; looking south, I can follow the coastline, scalloped by beaches and headlands; looking west, I see the forested hills rising up to the mountains; to the north is a grove of banksia below a burgeoning suburb; to the east, the rich blue of the Tasman Sea and the Solitary Islands. Just twenty meters beyond the information board on the paved pathway, I encounter a young kangaroo. He must have been lying in the grass because his fur is wet with dew. Ignoring the instructions on the sign, I continue walking along the path, my camera at the ready. I expect he will bound off to a safe distance or at least move off the path. Were this an encounter in the rangelands, he would have bolted when I came within fifty meters. But this is a suburban kangaroo. He leans close to me and sticks his nose into my camera lens—so close, in fact, that the camera can't focus. I take a photo, thank him, and go on my way as a couple of women in spandex and brightly colored sneakers march past. They are more curious about me and my camera than they are about the kangaroo. They will pass by another dozen before their morning walk is finished.

Given the right circumstances, kangaroos are incredibly tame. I've seen them grazing passively in front of campground kiosks, hopping up to nibble on sprigs of coastal heather proffered in the hands of small children, and accepting milk from anyone willing to hold out a bottle. I've seen the ease with which they grow up in suburban backyards, and how much they depend on humans to help integrate them back into the bush. The contrasting skittishness of the kangaroos I've seen in the rangelands, which is making kangaroo shooters' jobs increasingly difficult, is in no way due to some innate character of kangaroos. It's a response to hunting pressure. The theory proposed by Jared Diamond—that particular animal species (such as zebras) were never domesticated because of some innate intractability—can be thrown out the window in the case of kangaroos. They tame down as easily as emus, who lend themselves readily to intensive farming, as I've shown. In fact, kangaroos fit the bill in almost every characteristic that Melinda Zeder lists as behaviorally "pre-adaptive" to domestication.[34] There is nothing intrinsic to kangaroos in terms of development or behavior that inhibits taming and by extension farming, and this fact was not lost on the researchers who gave it serious consideration a few years ago. But is it true that the only inhibition to farming kangaroos is price? If kangaroo meat earned farmers enough money to justify the infrastructure and labor that kangaroo farming or ranching demands, does it follow that farming would be successful? From where I stand, a lot of other factors seem to inhibit kangaroo farming, and they speak volumes about the unmaking of Australia. One of these factors pertains to ownership. While common law in Australia allows landholders to hunt and kill wildlife on their land, some statutes and regulatory instruments assert Crown ownership of all wildlife.[35] These contrary positions are reconciled with a system of licenses and permits for killing wildlife, which makes for bureaucratic hoops that are nonexistent in the case of introduced animals such as sheep and goats. But the problem goes deeper than extra paperwork. Underlying the legal contradictions is the ingrained concept that animals farmed must be animals owned. Yet Aboriginal ways of farming kangaroos show that owning animals at the individual level is only one way of farming. For Aboriginal people, it is the land that is farmed. The animals they harvest are a consequence of maintaining the integrity of the land. This strikes me as a great motivation for maintaining healthy ecologies, but within the contemporary cultural context, is it viable? At present, the ownership of goats is pretty fluid because inadequate fencing allows goats to roam and select the best land. Farmers can muster only the goats within their property boundaries. But this arrangement is pretty precarious, and farmers are already moving to assert ownership over individual goats. In terms of farming kangaroos, the desire to own particular animals leads to the expectation that expensive fencing is required. This is in large part due to the perception that kangaroos are landscape floozies—that without fencing they will go wherever rain has recently fallen. While there might be some truth in this, research using GPS collars has shown that kangaroos have what biologists call a high degree of site fidelity. Apparently, they don't wander very far at all unless there's extreme drought. In the 1980s, Dale and Yvette McCullough used radio collars to track the movements and home range sizes of twenty-eight kangaroos of three species in Yathong Nature Reserve, NSW. The motivation for their study was to test the Australian folk wisdom that kangaroos are, like emus, homeless itinerants, following rain bands and evading drought. They found a species difference, reds displaying the greatest ranging behavior, western greys the least, and eastern greys somewhere in between. But all three species showed a marked fidelity to their home ranges. Males sometimes traveled considerable distances, but these were one-off trips that reflected a size, sex, and age class common in most mammals who are prone to wanderlust: young mature males. As for the majority, mean distances traveled within home ranges were very small. Over an eight-month period, these ranged from 0.6 kilometers for female western greys to 1.34 kilometers for male reds. The maximum distance was only 21.1 kilometers, and it was traveled by a female red whom they described as "a notorious gadabout."[36] For a region with an annual rainfall of only three hundred millimeters (about twelve inches), these distances are remarkably low. The conclusion: "No kangaroo in this study could be classified as nomadic."[37] In fact, the roos stayed within their home ranges even when thunderstorms could clearly be seen on the horizon. There is another telling study done by the National Parks and Wildlife Service in which researchers translocated four red kangaroos in Kinchega National Park, NSW. The kangaroos were captured, radio-collared, and moved thirteen to twenty kilometers from their point of capture, beyond the boundaries of their usual home ranges. Two of the roos, one male and one female, returned to their point of capture within ten weeks, and another male returned within one year. The fourth, a female, left her point of release and was moving in the direction of her former home range but was impeded by a 1.8-meter fence and never made it home.[38] Even these red kangaroos—notorious nomads—showed an enduring attachment to particular places, despite available resources in other parts of the park. As for grey kangaroos, in places with higher rainfall, their home ranges are tiny. A study from the Australian Capital Territory found that grey kangaroos confine themselves to ranges no larger than seventy hectares. What's more, the home-range sizes were influenced not by resource availability but by the presence of other kangaroos.[39] Still, long-distance movements of kangaroos have been recorded—in one case, more than three hundred kilometers across state lines.[40] But these are the exceptions to the general rule that kangaroos form attachments to particular places and only get itchy feet where the ground really dries out. This fact in itself does not rule out the feasibility of farming kangaroos; stocking rates for sheep in the rangelands are also determined by available resources, and sheep are farmed profitably. If no pasture is available in a given paddock, a farmer is not going to bring in a couple of thousand sheep. But in a cultural context that demands that every head of livestock be counted and that fosters unlimited accumulation of wealth, shared access to a movable resource would probably lead to overexploitation: whoever was fortunate enough to have a mob of kangaroos congregate on his land would take the opportunity to kill as many as could be processed and frozen.

This dynamic weighs on my mind after my close encounter with the young kangaroo on the headland. There's something deeply wrong with that encounter, and it's not just that I disregarded the instructions on the information board. While visiting Bourke, I walked into the butcher shop and asked if they had any kangaroo meat. The butcher laughed at me and suggested that I simply pick up some roadkill and save my money. Despite the abundance of kangaroos around Bourke and the widespread nightly shooting, there is not a morsel of kangaroo meat to be found in the stores—not in the butcher shops, not in the supermarkets, not in the pubs or restaurants. Down the road in Moree, there is a restaurant with goat on the menu, but no kangaroo.[41] I don't know why this is the case. Perhaps kangaroos are seen as pests rather than food, or perhaps locals need not buy the meat when they can kill it themselves. Whatever the case, the kangaroos being killed here all seem to be ending up somewhere other than the rangelands of New South Wales. In Woolgoolga, hermetically sealed kangaroo is readily available at the supermarket and butcher shops, and some of the trendier restaurants will serve up a kangaroo steak on a bed of bush-plum sauce or in a pot pie. I can frolic with kangaroos at the headland in the morning and have a kangaroo steak for dinner at a local restaurant. Late modernity has created profound separations not just between animals and complex ecologies but between people and the animals who serve as their food. It separates people from production systems and turns environments into burdens for farmers, and abstract concepts—or worse, playgrounds—for city dwellers. It spews tons of carbon dioxide into the atmosphere in a faraway place and sequesters it in a region where fire would otherwise make for healthy land. Within this paradigm, sustainable kangaroo farming is not practicable, because it requires a deep engagement with animals, land, grass, fire, and weather systems. It requires social networks of obligation and kinship and the self-regulation of a shared resource based on need rather than market trends. And it requires a quantum ontological shift from seeing land as measurable units of ownership to understanding it as a complex system within which kangaroos are sometimes integral but always gloriously ephemeral, fickle grass-tasting judges who only visit those willing to take responsibility for the health of Country.

ISOLATOR # 14

7.

Borderlands

It could just be possible to build an Australia that has as its beginnings not the invasion of this country, but the recognition that we are all connected. To the Earth. To each other.

—RICHARD FLANAGAN, "Everyone Suffers in the Politics of Hate"

When Jared Diamond theorized about why only Eurasian peoples domesticated animals and colonized the New World, he reasoned that tameness is crucial to domestication and that the animals available to Eurasian people were already halfway tame. He argues that this is largely why among the many sub-Saharan ungulates available to African populations, none were domesticated. Unlike the ancestors of our contemporary sheep and goats, gazelles tend to run at the slightest disturbance and keep on running. Meanwhile, goats and sheep simply mob up and move a short distance to higher ground—if it's available. On the flip side of the tameness coin is lack of aggression. Diamond compares asses (the ancestors of donkeys) with zebras and onagers and shows that, despite genetic relatedness to the point of being able to interbreed, the key difference among these three species, and the difference that led to only asses' being domesticated, is that zebras and onagers are incurably aggressive. Both species have a tendency to bite and not let go. In the case of onagers, who were extant in the very region where animal domestication took hold and were subject to fruitless domestication efforts by the ancients, the only reason we can find for their not being domesticated is their irascible nature.[1] The same emphasis on the importance of tameness was the guiding premise behind Belyaev's Siberian farm-fox experiment. The foxes selected for breeding were those who showed either little aggression or reduced fear toward their human handlers. Dmitri Belyaev believed that selection for tameness would elicit traits in farm foxes consistent with those of other domesticated animals. Again, the thinking was that the key to domestication was the capacity of at least some individuals of a species to be socialized to a human community.

The domestication of Australia's fauna shows us that tameness and socialization to humans need not figure prominently as key criteria for the animal species concerned. Instead, the criterion is that these animals thrive within a context of separation from their traditional ecologies. Dingoes' reputation for intractability, their tendency to escape, and their aggression toward other dogs all suggest that they are not good candidates for domestication—especially when other, similar-looking, inexpensive dogs are available. But dingoes' resistance to domestication is made up for with separations. High fences in suburban yards, muzzles, and leashes can overcome all the recalcitrance in the world, and technical fixes can temper aggression or flightiness. Dingoes can be separated from the bush whether they like it or not. As for stingless bees, they are already more amenable to humans than their distant European cousins, so the question of tameness is academic. The key factor will be the bees' ability to persist in urban environments and painted boxes. With human intervention, this certainly seems likely, and again it will be separation from pests, predators, and the elements that will define the domestication of stingless bees. When it comes to farmed crocodiles, tameness will in no way be crucial to their domestication. Farmers do not select against crocodiles who are aggressive toward humans; they simply learn to manage them. Though tamer animals might experience less stress, aggression is linked to growth rate that is selected *for*, so there is no simple equation for deciding which kind of temperament will thrive on a crocodile farm. Tame, aggressive, and diffident crocodiles will all be accommodated as new ways are found to separate them to the degree that they can be kept in single pens with very little intervention from humans. Meanwhile, tame emus are in fact selected against, because they attach themselves to farmers and won't let other birds near them. A degree of flightiness will be cultivated in farmed emus, ironically, to make them easier to manage. What will define farmed emus instead will be interventions to protect their eggs and chicks from climate and predators, and the separation of age classes from one another and of the entire farm population from unfenced landscapes. And as for kangaroos, their extreme tameness in association with humans is self-evident, but it hasn't influenced moves to domesticate them. I've seen wild roos hop up to small children and had one fog up my camera lens. They don't show any of the flightiness or aggressiveness that should theoretically prevent their being domesticated. The context of their farming will determine whether they become domesticated. In all of these cases, tameness, or the lack of it, in no way determines the outcome of processes of domestication. The common thread among all of them is separation.

While it builds on the same separations as domestication of old, the new wave of domestication in Australia is marked by a cultural context very different from the Neolithic, when people first corralled wild goats or brought home their wild kids. The early domesticators didn't have a working model of farming. They were following the food to its source, and their methods of making food more predictable evolved out of traditional subsistence.[2] For Neolithic herders, separating animals from their ecologies was not driven by the imperatives of capital. They had no money. The animals they herded may have had spiritual or totemic associations, but we can be pretty certain that in the case of livestock animals, it was all about putting food on the table, or whatever surface they were eating off. Behind this may have been other factors, such as growing populations or increased sedentism, but proximally it was all about food, and in some cases pets. We can also be pretty sure that these people had no preconceived ideas about how they wanted their domesticates to evolve. They weren't consciously selecting for behaviors or phenotypes based on a deep knowledge of heredity, nor were they aware of the potential consequences of what they were doing. In the words of Helen Leach, "Domestication was a process initiated by people who had not the slightest idea that its alliance with agriculture would change the face of their planet almost as drastically as an ice age, lead to nearly as many extinctions as an asteroid impact, revolutionize the lives of all subsequent human generations, and cause a demographic explosion in the elite group of organisms caught up in the process."[3]

The same cannot be said for us here and now. We have the benefit of being able to see in both directions. We can see into the past and understand the consequences of domestication for the animal species involved, the behavioral and morphological changes and the impacts on wild progenitors. Looking forward, we not only have the ability to imagine what we would like domesticated creatures to be, but we have the knowledge and means to turn these imaginings into realities. We can insert segments of DNA into the chromosomes of unrelated species to certain ends. But this godlike capacity doesn't set us above the rest of the world. We are deeply enmeshed in cultural frameworks that guide our decisions and set limits on what we might imagine.

In molecular terms, we are getting a very clear picture of the origins of dingoes and their population history on the Australian continent.[4] Archaeologically, the picture is not as clear. We can establish a latest date for the arrival of dingoes, but the exact timing and circumstances remain elusive. As to where dingoes are headed, this will depend on context. One hundred years ago, when nativeness was not as fashionable, colonial Australians were content to see dingoes go the way of thylacines. That contentment turned to intent and to programs of eradication. But since federation, and since the veneration of natives has come into fashion, dingoes have become icons of the Australian bush. They have gained a place in the country's natural history to the extent that people worry that they may become extinct by breeding with nonnative dogs. The solution to this perceived problem is to separate them from the bush that defines them and treat them like a breed of dog with a wild streak, to maintain the purity of nativeness, something that can't be done with bush dingoes, who will breed with any old mutt. But aside from econationalism and panic over hybridity, other forces are also at play. People are fond of dogs. If a species of frog were threatened with hybridity, there wouldn't be the same panic and public determination to preserve the frogs' pure bloodlines. But dingoes are as close to dogs as wild canids can be, and their puppies are especially cute. They elicit nurture and passionate love that translates not only into caring but into wanting one around the house. That Australia was colonized by a dog-loving nation ensured that dingoes would make for much-loved pets rather than savored delicacies. This is why dingo domestication is proceeding without a profit motive. Apart from field biologists, I haven't come across anyone who is making money off dingoes. In fact, most dingo lovers seem to be looking for ways to raise money to feed their charges. It is love, not capital, that's driving dingo domestication.

Stingless bees are being domesticated within a similar set of reference points, only without the panic over hybridity. Their nativeness is as endearing as their stinglessness; people are applauded for rescuing them from meter boxes and taking them home to be kept in the backyard. Recall the smile and thumbs-up I was given when I scooped the brood from the fallen tree on Boggy Creek Road. I was doing my patriotic duty, rescuing legitimate denizens of the Aussie bush. That I was taking them home in a box to the suburbs where they would pollinate introduced plants was not a concern. As with dingoes, a native is defined by what it is rather than what it does. Within this context, those who domesticate native bees are applauded even if they live in a climate so cold they have to wrap the hive in an electric blanket for the winter. And, as with dingoes, there is not a lot of money to be made from bees. Some folks make a living off them, but nobody is getting rich. In fact, there is no profit motive at work in producing resin or honey; the money to be made comes from selling hives and equipment, hiring hives out for pollination, and giving talks about stingless bees. Stingless bees are primarily kept for enjoyment, so they almost qualify as invertebrate pets. In this way, stingless bee domestication aligns with Euro-Australian ideas about ownership. Regardless of where the bees are going or what they're feeding on, ownership operates at the hive level and everything else is incidental. We might take this as a given in terms of domestication, but it rests on a particular way of conceiving of the world. It's a major departure from the Yolngu way of keeping bees, which rests on attachment to place; hives are a consequence of that attachment. We are so entrenched in our view of hives as separate units of production that we find it difficult to imagine an alternative.

Econationalism doesn't underpin crocodile domestication to the same degree that it does dingoes and bees. This is partly because saltwater crocodiles are not unique to Australia. They are accomplished travelers of ocean currents, and for the time being their range extends from the east coast of India to Vanuatu in the Pacific. What aligns crocodiles, bees, and dingoes is conservation, and this is what constituted the rationale for the first crocodile farms. The need to conserve crocodiles emerged out of a particular cultural framework not limited to the singularization and preservation of species. It was also colonization and the global flow of capital and goods that made wild crocodiles into valuable commodities; thus they were hunted to the extent that conservation became necessary. Now it is those same global flows that sustain crocodile farming; there is no market anywhere near Bindara for fifty-thousand-dollar handbags. While the vagaries of the fashion industry might augur a precarious existence for producers, crocodile skin is to the industry as canvas is to painting: fashions may change, but crocodile skin will presumably continue to constitute a material from which changing fashions are made. This use of crocodiles gives rise to very clear ideas about what domesticated crocodiles should be like. They should produce lots of eggs and lots of belly scales with a minimum of fuss in terms of feeding, housing, breeding, and management. This in turn is why artificial insemination is so attractive to crocodile farmers: because it will allow them to create the kinds of ecological separations needed to produce valuable skins. So the future of crocodiles seems to lie in their genetic material, which must be manipulated to overcome their (anti)sociality. The people at Bindara insisted that it was impossible to farm crocodiles without a love for crocodiles, but whether this love will persist through the new modes of farming remains to be seen.

As for emus, I can see how love need never enter into the equation. Shane the emu farmer is not particularly passionate about his birds, at least not in the same way as Terry Walker, whose much-loved birds are on the way to local extinction. This is not to say that a love of emus is incompatible with emu farming—just that it doesn't seem to be as crucial as it currently is in the case of crocodile farming. Shane's passion is not for the emus but for their oil and its therapeutic properties. Aboriginal people were well aware of these properties, but that didn't inspire them to domesticate emus, despite the birds' tractability. It takes a particular cultural context to envision a paddock full of emus being transformed into marketable bottles and capsules of oil. But the scalability of emu farming is not simply a factor of the efficacy of the oil. Emu farming only works in a modern world of global capital, where the same forces that make the oil profitable also, ironically, contribute to the kinds of diet-related health problems—high cholesterol and diabetes—that the oil is meant to treat. Given that these health problems will not go away anytime soon, the future of emu farming will be similar to crocodile farming. Already separated from migratory routes, predators, cosmologies, and one another, emus will be unmade in ways yet to be imagined. Artificial insemination will remove the need for mated pairs to apportion paddocks, and female-only paddocks will be able to hold denser concentrations of birds. I would be very surprised if the farmed emus of two hundred years hence are not markedly different from their wild cousins.

It is the same cultural context that has so far prevented kangaroos from being farmed. I've shown how they could conceivably be farmed, but this would require a paradigm shift not just in the definition of farming but in its practice. And that in turn would require a shift in the mindset of colonial Australians who have difficulty accepting kangaroo meat as food. Kangaroos were a staple of Aboriginal diets in the outback, but colonial Australians see them differently. They are indeed native, and this lends cachet to kangaroo meat as ecologically sound, but there's a national inhibition about chewing on meat that wasn't onboard the first fleet of ships that arrived in 1788. Perhaps in time this inhibition will be overcome, but even so the question remains whether a strong market for kangaroo products will translate into farming, given the constraints of the Euro-Australian model of raising livestock. Potential kangaroo farmers' imaginations extend only as far as ownership of mobs of kangaroos in fenced enclosures. If these conditions of separation lead to disease, as with the intensive farming experiment of the 1970s, then farming methods will demand veterinary treatment of kangaroos and more separations. There is nothing intrinsic to kangaroos that precludes farming—the stumbling block is that colonial ways of farming preclude kangaroos. Not that this is a bad thing, particularly if you're a kangaroo.

These new wave domesticates—dingoes, bees, crocs, and emus—are being separated from their ecological niches and thus reduced. Kangaroos are not, primarily because they resist. In each case, the Euro-Australian colonial cultural paradigm sets limits on what is possible in terms of both imagination and practice. Dualisms permeate the process. The same frames of reference that foster the impossibly discrete notions of good and evil lend themselves to unmaking animals through farming and pet keeping. Conceptions of *native* versus *introduced* underpin the validity of these domestication projects in Australia, although in the case of dingoes they precipitate a lot of ontological hand-wringing. Here, the dualism of pure/hybrid lends urgency to the separation of dingoes from their ecologies and reiterates the nature/culture dualism that also demands separations and pits farmers against pests, predators, and the elements. In making these separations, civilization pits itself against wildness, and reason against emotion. From there, the only direction available is reduction and isolation—the reduction of domesticates to measurable units that have been extracted from their ecological contexts and considered in isolation from everything else. Unmaking.

I have a sister in Melbourne who works for a pharmaceutical company. It's not an ordinary pharmaceutical company—it makes vaccines for chickens. These birds have come a long way from the forests of Southeast Asia. Having been domesticated eight thousand years ago, they were being bred in Syria and China by 4000 B.C., and in the hands of people they had reached Africa and western Europe by 1500 B.C.[5] In terms of domestication, chickens are an emphatic success story. At more than twenty billion individuals, they are not only the most abundant of the world's domesticated animals but the most abundant bird species on the planet. They also represent the apotheosis of unmaking. The development of new vaccines has been necessary to protect farmed chickens from a host of pathogens. Not all are used on any given farm—particular conditions dictate which vaccines are appropriate—but their purpose is to isolate chickens from the pathogens that cause disease. The company my sister works for is actively seeking to steer farmers away from antibiotics, which have traditionally been administered as a prophylaxis rather than a treatment. If not for World Health Organization guidelines, government regulations, and increasing consumer concerns, chicken meat would be rife with antibiotics.[6] While there is increasing demand for free-range chicken meat, conventional chicken production is all about separating chickens from everything else. Chicken farms should be located at least five kilometers apart, with two-meter-high security fences and controlled gates, effectively quarantining each farm. The chickens are kept in vast sheds within which the temperature, humidity, and air flow are controlled to create the optimal conditions for growth and health. Other than humans, industrially farmed chickens have no predators—no foxes, no hawks. Pests such as beetles and flies are controlled with insecticides. In addition to vaccines and antibiotics, pathogens are controlled through hygiene practices. Visiting vehicles pass through a disinfectant wash; visiting humans go through a disinfectant shower/change facility. Prior to entering sheds, people pass through a disinfectant footbath, and movement between sheds is discouraged. Used equipment is not taken onto a farm as it can be a disease vector. Sheds are cleaned and sanitized every six weeks. Feed consists of cereal carbohydrate and vegetable and animal protein, bare nutrients that have been separated out from their sources. Even photons have been separated out on chicken farms; red light increases the incidence of pecking, so the optimum is blue and green light or dim lighting. The ideal of chicken farming is to isolate chickens from uncontrollable environmental conditions and all forms of organic life, even to the point of employing robotic technologies and thus excluding even humans from chicken management.

Chicken farming represents cutting-edge domestication, where the creation of an ecological desert occupied by chickens alone is the logical end point. This model may be the vanguard for the new wave of domestication; we must acknowledge that the same degrees of separation might be realized for the native Australian species currently being domesticated. But is there an alternative? Traditional ways of engaging with animals in complex ecologies might sound ecologically sound, but are they at all practicable in a world of nearly eight billion humans? Perhaps we've populated ourselves into a corner, leaving ourselves no choice but to create ecological deserts such as chicken farms and continually increasing measures toward ultimate separation. There is little space for biodiversity in a high-rise apartment in Hong Kong; a call-center worker in Delhi has no time to wander around the forest knocking on trees, looking for sugarbag. This is a question worth asking; we cannot deprive millions of humans just because we want our chickens to have access to grass.

Many people argue that we need to move to a plant-based diet. Aside from the profound cultural shift required, in terms of ecological deserts this promises the same end point as animal husbandry. The world's human population demands such a vast quantity of food that monocultures seem to be the only way forward, and they too are all about separations: isolating crops from ecologies to the point of genetic modification, where target plants are engineered to resist herbicides and pesticides that kill everything else. Ecologist Paul Shepard tells us how the predominance of vast monocultures of supercrops is creating balances that can't be kept.[7] He has a point. The organisms that these separations are supposed to defeat do not go away. They evolve and adapt, ratcheting up the need for more technological inputs and interventions, while the shadow of the human population grows longer. Meanwhile, the dictates of global capital allow us to take something from one place to prop up the deficiencies in another, as though we could fiddle with the pieces without consequences. Unmaking might serve us in the short term, but its legacy will be massive degradation and pollution of soil and water that once sustained us.[8] Of course, the consequences will not be distributed evenly, and those who benefit most from unmaking will, ironically, be the ones least affected—in the short term. But the dogma of individualism, isolation, and separation must inevitably fail. Even the best-built walls cannot disrupt the connectedness of the earth's systems and the interdependence of its organisms. By hook or by crook, the great unmaking will in the end impact the entire human population, causing radical changes in the ways societies are organized. The remaining question is whether we choose these changes or whether they are thrust upon us.

Assuming that we are still in a position to choose, what might an alternative look like? Here I return to Debbie Rose's Indigenous philosophical ecologies, to see whether the concept might lead to some practicable alternatives to unmaking, some directions that lead away from separations. For the sake of broader appeal, I'll refer to the concept using a more accessible term—*reconnection*—and address each of the four points of dialogue as steps toward reconnection, although referring to them as steps does not imply that they should follow any particular order or that they preclude any dialogue. The first step is to extend subjectivity beyond humans, recognizing sentience and agency among all participants in life processes. This is the polar opposite of factory farming, where subjectivity is subsumed by scale, process, and profit, where "nothing is more worthless than an individual chicken."[9] At the same time, for practical reasons, this move must sidestep animal rights discourse, despite the obvious impact that industrial farming has on animal welfare. This is because extending subjectivity beyond humans within an ecological framework entails undermining the basis of human dominance over animals. I am not saying that subjectivity across species is inconsistent with animal rights, as defined by the animal rights movement, but that it diverges from the anthropocentricism that in many cases forms the foundation for animal rights discourse. We need to recognize that the concept of rights, and indeed of sentience, within a philosophy of reconnection is different from that of animal rights discourse, which demands a hands-off and/or a paternalistic approach on the part of humans and generally accords rights to nonhuman animals based on things like sentience or the capacity to feel pain, rather than on their integration into entire ecosystems. Within a philosophy of reconnection, humans are not arbiters or benchmarks; rather, we are participants, with obligations toward other organisms based not on their similarity to us but on their participation in the same life processes as us. I've described how crocodile agency has impacts on the affairs of humans. In terms of conservation, the agency of crocodiles challenges the human-centered ideology of mastery over nonhuman nature. Modern crocodile conservation must in many respects overcome the subjectivities of crocodiles; they must be abstracted and subsumed within a numbers discourse. The same applies to a farming context, but in a farmed landscape the subjectivities of crocodiles are much more in your face. They cannot be ignored, because these toothy five-hundred-kilogram reptiles' agency affects the farm's bottom line. Their demands for particular farming practices and particular levels of attention are at odds with what a late modern farm aspires to be, so separation and abstraction become logical consequences of crocodiles' subjectivities. Yet these subjectivities are not going away, and to reconcile them with farming would require a paradigm shift in terms of farming practices—which would in turn necessitate a paradigm shift in the global flow of money and crocodile skins that demands separations at the farming level. This is a serious stumbling block. We seem to be faced with an either/or situation: either the structures of late modernity must be dismantled and reconfigured, or the subjectivities of other organisms and ecosystems must be dissipated into abstraction. To be realistic, this doesn't bode well for crocodiles as subjects.

The second step toward reconnection is displacing humans from the apex of food webs and resituating them in local patterns of connectedness. This is no small ask in a society that commoditizes chickens for the sole benefit of human consumers. In terms of domestication, a nonhierarchical approach might seem unachievable, if not downright absurd. Recall the definitions of domestication that include concepts such as mastery, control, and ownership. These are not particularly egalitarian concepts. But recall too the worldview of Mongolian herders, who situate themselves in mutualistic relationships with their herd animals. When they say "We feed them and they feed us," they are attesting to a reciprocal relationship in which domestication is a mutual process within a hybrid community of herders, livestock, and sometimes predators. There are separations—livestock animals are protected from the elements and from wolves—but these are in part due to choices made by the animals themselves. And the wolves are accorded status not as vermin but as ancestral spirit animals, as integral to herding landscapes as herders and herd animals.[10] So human hegemony is not necessarily a default position with regard to farmed animals, but the question remains: once unmaking is ingrained in a worldview, can it be undone? How can we see ourselves as integrated into food chains not just as consumers but as consumed—as "part of the feast in a chain of reciprocity"?[11] Matthew Hall proposes "ecological anarchism" as a means by which to resituate humans within ecologies. Traditional anarchism not only rejects the imposition of state governance; it rejects the imposition of power relations and seeks to delegitimize social hierarchies. Historically, these hierarchies have been made anthropocentric by allowing for dominance over animals and plants for the benefit of humans. But Hall proposes the extension of anarchism to include ecologies. Hall suggests that egalitarian "tribal" societies and the "distinctly ecological" outlooks of Indigenous people point the way forward, allowing us to relate to ecologies from within rather than from above.[12] What's more, this perspective does not necessarily entail a rejection of agriculture or domestication, because from within ecologies, domesticates can be perceived as collaborators in the activities of an ecoanarchist movement in much the same way that Mongolian herders perceive their herds.

There is nothing pernicious in this proposal; indeed, egalitarian and state-free is how human societies have functioned for most of human evolution. But aside from anarchism's obvious branding problems, it works far better as a concept than it does on the ground. The tribal societies that Hall refers to could indeed be exemplars of ecoanarchism, and in many cases they demonstrate that an egalitarian human society is highly compatible with a nonhierarchical relationship to other animals and plants. And, yes, such societies have great track records of engaging with the environment in ecologically sound ways. But the problem with Hall's argument is that, historically, tribal societies have a poor survival rate when confronted by hierarchically organized, state-based societies. So the success of ecological anarchism would depend on the absence of state-based societies. Not only would ecological anarchism be difficult to enforce without state intervention, but preventing the emergence of states within an agricultural context would be akin to a game of anarchist whack-a-mole. History shows that, regardless of time or place, wherever there is agriculture, social and economic hierarchies inevitably pop up, and they don't pop back down without an ecological crisis, and this is the heart of the problem with resituating and reconnecting humans within ecologies: it cannot be achieved without undoing hierarchies within human groups. Val Plumwood has argued quite convincingly that ecological rationality cannot be achieved without social equity, and that this cannot be achieved within a capital-driven liberal democracy. The inequalities that inevitably emerge from such a society create different degrees of remoteness from ecologies and ecological harms. Plumwood points to the ways in which the environmental costs of extraction, production, and consumption by those at the top of the social spectrum are typically borne by those at the bottom. This hyperseparation automatically confers a dangerous apex position with regard to ecologies, because the perception of the privileged is that ecological harms will primarily affect the poor. So as long as society is hierarchically organized, those at the top will be incapable of positioning themselves within food webs and local patterns of connectedness with the nonhuman world, because there will always be others lower down who are positioned closer. For the privileged, there can be no connection to local ecologies because, for the time being, at least, the entire world is available, whether to exploit or in which to find a place to hide—or both.

The third step toward reconnection extends kinship and morality across species lines. Paul Shepard's prescription for an ecologically healthy society calls for the establishment of associations with responsibility for particular species.[13] These amateur associations would study and monitor their totem species, and any kind of development would need to be approved by the associations responsible for the species affected. But this approach is more about totemism than it is about kinship. For many Aboriginal Australians, totemism is a "blunt instrument" that fails to capture the immensely complex relationships between humans, animals, plants, and features of the landscape.[14] It falls short in suggesting a metaphorical or representational relationship, rather than kinship, with a totem species. When a Yolngu woman says she *is* crocodile, she's not saying the same thing that a member of the Townsville basketball team is saying. She is asserting that she is crocodile in the transubstantial sense that both she and the crocodile trace their ancestry to the same progenitor, who exists across dreaming and modern landscapes and confers certain obligations on both human and crocodile. In this sense, kinship operates nonhierarchically and provides space for morality across species lines. Yet the moral obligation toward crocodiles is not limited (as it is in the totemic sense) to care only for the animals emblematic of the species. As Debbie Rose shows us in her accounts of Yuin and Ngyamba (Ngiyampaa) people, kinship with other species in this context confers a duty of care that inheres in bonds of mutual life-giving and includes care for any elements of the environment that enable the totemic species to thrive. In other words, sharing ancestry with a crocodile does not merely entail respect for crocodiles and perhaps a prohibition against harming them. Rather, it demands that a person consider a crocodile's every connection to the surrounding ecology and ensure that these connections are kept intact so that crocodiles might flourish. It requires a consideration of the rivers and tributaries and vegetation, and the landforms that feed the tributaries, and the animal species that affect the vegetation, and the plants and animals that feed those animals. These relationships are in turn amplified as kinship extends beyond single species—thus an individual might have associations with four or more significant species, all situated within overlapping complex ecologies.

This is the elegance of kinship in Aboriginal Australian terms: that it not only recognizes our shared ancestry with other species but calls on us to see other plants and animals in terms of their entire ecologies. It demands attention to complexity and confers responsibility beyond particular plants and animals in particular places. This responsibility then shapes traditions and institutions: in prohibitions against overexploitation, timing burning regimes, protections against ecological damage, increase ceremonies, and so on. How would these traditions translate into a pet-keeping or farming context? In the case of contemporary dingo domestication, they don't. No matter how much someone loves a dingo and incorporates her into a human family, as long as there are 2.4-meter fences, flea collars, and leashes, the kind of kinship demanded by reconnection is just not possible. There's a certain kind of instrumentality in keeping dingoes that rejects their embeddedness in Australian ecologies. While they're ostensibly being bred for future wild release, this serves a culturally mediated, subjective desire for genetic purity rather than a desire to nurture dingo ecologies, which now include Euro-Australian dogs. Certainly, it serves a similar desire to nurture, and a love for dingoes that also motivated Aboriginal Australians to take dingo puppies from their dens. But the key difference is in the positioning of this pet keeping within ecologies. For Aboriginal Australians, engagement with dingo pups occurred within a context of connectedness that allowed adult dingoes to disperse from their adopted human groups. Both dingoes and humans maintained their connections with the bush within this system. For colonial Australians, engagement with dingo pups entails separating them from the bush.

The same kind of instrumentality precludes kinship with other farmed animals. For example, it is not necessarily problematic to collect and incubate crocodile eggs in season, although it is a first step toward unmaking them. What does preclude reconnection is the commoditization of crocodiles, which doesn't necessarily objectify them but severs connections that are integral to kinship. Raising a crocodile on a farm not only undermines the crocodile's importance to food webs and crocodile populations within river systems; the commoditization of the crocodile's body in the global market takes that life out of the system and sends it somewhere else. In such cases, a life is not returned to the ecology to nurture life; instead, something is taken away. In terms of the animals discussed in this book, backyard beekeeping comes closest to the concept of kinship within reconnection. Bees rely on local ecologies, and their honey and propolis can be used locally. But modern beekeeping fails ecologies in its antagonism toward syrphid flies, phorid flies, hive beetles, and other creatures who, despite the harm they might cause to hives, are still integral to the system. Within the concept of reconnection, there is a place for pests and predators.

The fourth step is to challenge human autonomy and allow other species to call us to action. What this refers to is Aboriginal people's responsiveness to the cues and demands of their ecologies. This is yet another means by which to deconstruct hierarchies and situate humans as participants within food webs. For example, humans must respond to seasons; Debbie Rose shows how seasonal calls to action are encoded in ecological patterns. When the March flies are biting in the Victoria River region, the crocodiles are laying their eggs. It's a call to action on the part of the March flies, and the people respond by collecting and consuming the eggs. Or take the song of the cicadas. When the cicadas are singing, the figs are ripe. The cicadas call people to come and eat figs but also to throw some into the water to attract turtles, which are good to eat at that time of year.[15] Even burning times are dictated by the land, as the flowering of the silky oak signals the time to burn. At one level, these seasonal calls to action resemble the rhythm of farming, the activities of which are also determined by the seasons. When crocodiles lay their eggs on a farm, it is time to collect them and put them in incubators. When the weather is warm and there is a lot of forage, it is time to split beehives. But at another level, seasonal calls to action and farming are worlds apart. While the wider Australian system relies on linear time, calendars, and reminders sounding off on cell phones, the traditional system seeks out connections in radial time. It maps a network of flowerings, hatchings, matings, bitings, and singings onto a complex world that fosters interdependencies. There is no need for linear time in such a system, because the landscape doesn't present the world in a linear way. Rather, it presents patterns that engage participants in processes of mutual benefit.

Reconnection, in the way I've framed it here, seems like a poor fit for late modernity, and indeed it is. The deep-rooted institutions of the great unmaking resist ecological connections with the stubbornness of a rock hanging precariously over the tide. The prescriptions of an Indigenous philosophical ecology might be adapted to wilder places, and certainly to lands that remain under Aboriginal stewardship, but on the land that feeds the cities it is up against some powerful forces of separation. The kind of ecological meshwork that must be nurtured in a process of reconnection is anathema to domestication, which pulls in exactly the opposite direction. Domestication by its very nature cannot be reconciled with reconnection. But let us not throw the baby crocodile out with the tank water. Certainly, domestication is inherently ecologically unsound, but this does not automatically mean that farming is—though it may be necessary to rethink our conceptions of farming.

I've driven through an industrial estate to the northern suburbs of Brisbane, where broad streets and freestanding houses speak of a time when land was abundant and cheap. I'm visiting Tye Kennedy in his weatherboard home on a quarter-acre block. Tye has a prominent forehead, a long curly beard with streaks of gray, and gold-colored eyes, all of which give some indication of his ancestry: Yorta Yorta, Radjuri, and Tharawal, people of the Riverina and New South Wales coast. But he also has European ancestry and attachments to other places. Tye grew up in the suburbs of Perth, a continent away from the country of his Aboriginal ancestors. As he says—and not without some regret—he's not part of the "cultural clique" that defines less displaced Aboriginal people. He can't closely identify with people like the Yolngu, who have a relatively unbroken connection to their ancestors and their homeland. He can't walk through country and find meaning in every lizard and every stone under which they hide. Instead, he's stuck in the suburbs feeling a sense of loss. But Tye wants to own that loss and accept that part of himself that is not of this land. "I can't just discard that part of me that's not Aboriginal," he says. "I have to accept that stuff and move forwards. We all do." So Tye is searching for a way to connect his ancestry with the reality that is contemporary Australia and to do so with integrity. He and his friend Neil began an enterprise called Little Black Bee with this purpose in mind. They see native bees as a key to unlocking the potential of the Australian bush and its traditional custodians. They want the bees to show them connectivities that might point toward alternative ways of engaging with land. But they also recognize that this is a massive challenge, in terms not only of understanding the intricacies of the Australian bush but also of overcoming social and political structures that drag like an anchor on any way forward.

Tye is both contemplative and plainspoken. He responds to each of my questions with thoughtful silence before drilling his blunt truths into the table with his index finger: "The concept of farming in this country needs to fuckin' change. Those first colonial farmers were so close to doing something right with native plants and animals but they missed the mark. They were watching what the blackfellas were doing but they had the wrong attitude, and that paved the way for the modern way of thinking. So now, one hundred and fifty years down the track, farmers are only just beginning to realize that they can't keep going the way they are. I know an old farmer who's only now saying, 'Hey, maybe there is something to this burning.' It took him seventy fuckin' years to arrive at that point!" And it's not just a matter of taking bush foods and trying to wedge them into a colonial farming paradigm. "It's been tried," says Tye. "Some enterprising types in South Australia took Quandong trees and planted them in neat little rows. Quandong moths came and devastated the lot."[16] Any productive way forward, as far as Tye is concerned, needs to recognize pests as integral parts of ecologies. He shows me a complex network of spider webs at the bottom of the garden. "These are golden orb spiders. The thing with these is that they only showed up after we put the bee hives in the garden. The webs are right in middle of the bees' flight paths but they hardly ever catch any bees. So just maybe they're catching bee-flies. I've seen a bee rolling around in the web and the spiders doing nothing, so I'm thinking that maybe the bee is leaving pheromones in the web to attract bee-flies. But we just don't know, and we don't know what we don't know." It is these sorts of connectivities that Tye sees as the way forward. He and Neil want to buy up land, introduce bees, and plant systems of bush foods. But even the concept of buying up land is problematic within the paradigm they're seeking. "What we're up against is the brokenness of it all, and what's underlying it is this concept of what's mine is not yours. We've run with this concept of ownership and screwed this placed beyond all recognition, and we're stuck with a world and how do we make it better? Let's stop dickin' around with this, and let's acknowledge the problem and have some truth talks around it. Let's treat this as a conversation that people need to have now. There's no resistance on the side of the blackfellas. I don't see any Aboriginal people out there saying 'No, it's our intellectual property and piss off.' It's been totally the opposite. So where's the resistance? It's not on that side. It's here. It's with farmers and multinationals, governments and consumers. The truth could set ourselves free and open up some unbelievable things that could open up agriculture—open up Australia's identity to the rest of the world. We need to strengthen Aboriginal communities, then go and find what they're good at and triple down on that. That gives a way for the rest of us who are wanting to reconnect. It gives farmers a way to do what they do differently and better. But I don't see it coming from the top down. It's going to take a groundswell. Like green energy, it's got to start with people."

"What's mine is not yours." In 1974, anthropologist Norman Tindale published a map that was the result of fifty years of fieldwork across the entire continent of Australia. The map was intended to challenge the British imperial dogma that prior to colonization Aboriginal people were itinerant wanderers of the bush, a dogma that conveniently exonerated the colonists who invaded and occupied the land. The map, titled "Tribal Boundaries in Aboriginal Australia," depicts Australia as a mosaic of territories, each associated with its particular language group. Tindale's map certainly proves Aboriginal occupation of particular places across Australia, but the details have been a source of much contention. Debbie Rose and her colleagues note that "in depth research has almost invariably failed to confirm the information on his map in its entirety."[17] But regardless of the accuracy of the boundaries and the groups Tindale named, the map is interesting because it graphically depicts a particular worldview of land as polygons of ownership. Compare Tindale's map with a painting by senior Pitjantjatjara man Nyerri Morgan.[18] The vivid black, white, and red painting presents Morgan's primary estate, Karlaya, as a central node of concentric circles made of white dots and black lines. Radiating out from Karlaya are several other black lines (links), which join with other concentric circles representing other estates. While the painting is ostensibly a map portraying discrete estates, it stands in contrast to that of Tindale because it represents not boundaries but connections. The links represent the dreaming tracks of Morgan's ancestors, which in turn represent custodianship of other estates. Through initiation, Morgan's relationship to the land writ large is defined not by boundaries but by connections and responsibilities toward other people and other estates. Indeed, there can be boundaries between estates, as our indecisive crocodile friend demonstrated earlier, but these do not define the land. Rather, it is connections between places and their people that are primary. Boundaries in the colonial sense have no place here; they are the marks of control of land, of militarism, of possession, wealth, and commoditization, which stem from a particular way of being in the world: the way of unmaking. What is more, as Tye succinctly puts it, the concept of ownership of discrete territories represents a major obstacle to engaging with the land in an ecologically healthier way. When land is divided, owned, and fenced in the traditional colonial way, it becomes a means to an end rather than an end in itself. It is a constrained, instrumentalized canvas on which to paint domesticated plants and animals. The boundaries determine the extent of these organisms' ecologies and encourage the concentration of and care for single species to the exclusion of everything else. It's a marked contrast with traditional Aboriginal modes of occupation, which focus on key places as central to connected ecological nodes. This is in large part why fire-stick farming is inimical to unmaking: fire doesn't respect boundaries. Even if farmer Sue was willing to burn a particular paddock to make her land healthier, farmer Bob would throw a fit when the fire crossed his fence line. Ownership of land in the colonial sense inhibits a systemic approach to healthy ecologies; only in reducing the primacy of boundaries and enhancing communities can progress be made. Consider the success of Landcare Australia, the organization behind the meeting I attended at Fords Bridge. Landcare is the largest environmental-management movement in the country. Initially a government initiative, it is community owned and driven. There are now more than 5,400 Landcare and Coastcare groups, and the model has been adopted in more than twenty other countries. The success of the Landcare model is not based on respecting boundaries; rather, the model brings communities together, crosses boundaries, and treats land holistically. The premise of Landcare is that the results of land-management practices have no regard for fence lines. The model's viability depends on the recognition that land and water elicit shared responsibilities and a willingness to cooperate in creating sustainable agricultural practices and more diverse ecologies. There is nothing in the Landcare manifesto that explicitly says so, but boundaries defined by ownership constitute significant impediments to this end.

Where land is commoditized and ownership is cumulative, ecological disconnections multiply. This process operates in two ways. The first is that cumulative ownership of land fosters social inequality. When ownership is unimpeded by social controls, then ownership begets more and more land. It's also a fact that land is power. The cumulative concentration of ownership in fewer and fewer hands naturally leads to inequality and to the dependence of the landless on the landed. This in turn fosters the hierarchies of remoteness that Val Plumwood refers to, which situate people outside rather than within ecologies.

The second way that commoditization and cumulative ownership of land foster disconnection is by separating the landed from ecological harms and, by extension, from their moral responsibilities toward the land itself. When someone owns land but does not depend upon it, he has little incentive to ensure its health. Again, it is remoteness that inhibits ecological reconnection—in this case literally. Living somewhere else—other than on the land that one owns—depersonalizes that land and again undermines any incentive to maintain its health. We are so inured to the harms that arise from ownership, and so attached to the cumulative model, that we fail to realize how contrary they are to human history. Only in the last ten thousand years have humans been able to accumulate and control land to the exclusion of others; before that, land was something altogether different. In Australia, commoditization and cumulative ownership are only 230 years old. If we are serious about reconciling colonialism with traditional Aboriginal ways of life, then we must acknowledge the reality that this model of ownership cannot persist. Restricting ownership of land is anathema to neoliberalism, but only because we take land ownership for granted. Even the staunchest capitalists—at least most of them—would not countenance the commoditization and ownership of human beings today. So why not a similar prohibition regarding the land on which human beings depend? We can do it with ivory and rhinoceros horns—why not with land? Restricting land ownership would have two obvious benefits. It would reduce social and economic inequality, which give rise to all manner of social problems, such as crime, poor health, lack of education, and mental illness.[19] And it would make healthy land a common imperative, because healthy land creates attachments to local ecologies. Colonial Australians need to start seeing land as a moral subject and consign its commoditization and cumulative ownership, like slavery, to the trash heap of history.

As George Monbiot notes, "Most human endeavors, unless checked by public dissent, evolve into monocultures."[20] In terms of farming plants and animals, capitalism has a natural tendency to focus on whatever derives the most profit from a particular region, while at the same time excluding other organisms, whether harmful or not. It will take a supreme effort, and not a little sacrifice, to do things differently. Monbiot mourns the broken strands of life in domesticated landscapes, with their simplified ecologies. His solution is "rewilding"—expanding food webs vertically and horizontally, increasing the number of trophic levels (the positions that organisms occupy in a food chain), and increasing complexity in ecosystems.[21] Monbiot is right. We need to increase complexity in farmed landscapes, not reduce it. Like engineers, we need to build redundancy into ecosystems that will in turn make them more resilient. But we also need to recognize that systems don't have boundaries; they meld into neighboring systems where creatures, spores, nutrients, water, soil, sand, and weather systems create strands of connection that hold everything together. We need to allow and encourage these connecting strands, and that means deconstructing boundaries. Emus, for example, would need to be unfenced and unhindered in their roaming. Entire communities would need to farm them at the landscape level. They would need to be seen as a shared, ephemeral resource. As in the case of kangaroos, the infrastructure for the processing and use of emus would also need to be shared. Critics might argue that a resource shared in this way will soon be exhausted. But this is not inevitable. Fire-stick farming created mosaic habitats and an abundance of kangaroos, available for harvest by anyone with access to the land and not prevented by certain food taboos or clan restrictions. We know that the Aboriginal Australian system of farming worked, because European colonists in 1788 found it perfectly functional and intact after sixty-five thousand years in practice. What is required, though, is a community-level commitment to creating and subsisting on healthy, complex landscapes, and trust in the natural processes that sustain lives.

This is a key difference between Aboriginal and late modern ways of farming: in the latter, people are alienated from their means of subsistence. People are increasingly dependent on food produced thousands of kilometers away and shipped to them in plastic wrapping. Folks in Norway are eating mangoes and bananas, and folks in Chad are eating tinned sardines. Whereas free trade, fossil-fueled transport, and the internet are creating hitherto unimaginable connections between humans in an increasingly connected world, these processes are eroding the connections of people's trust in and dependence on what keeps them alive: the land. People gravitate to cities where high-rise apartments, networks of concrete, and sterile packaged foods ensure that they need never touch any kind of soil. City dwellers have no motivation to care for the land they live on because it is buried beneath concrete and permeated with pipelines. Their only concern is for an abstract *environment* that in practical terms is always somebody else's problem. The land beneath the hooves of the cattle from whom their steaks are cut is meaningless in such a world. This is why the neoliberal dream is an ecological nightmare: producers and consumers are not motivated to ensure the health of ecosystems. The producer/consumer relation will readily extract from a place until it is exhausted, because the whole world lies before it, and it can move on and exhaust one place after another.

By all means let us live off native animals, but not as late modern domesticators or consumers. Rather, we need to see ourselves as caught up with these creatures, connected to them, in local food webs for which we are not only responsible but within which we belong—as both feeders and food. Rather than take inane photos of kangaroos for real estate ads as a salve to psychological emptiness, we should be killing and eating them where we live. Of course, this suggestion will cause either revulsion or censure from Westerners, because ecological separation is so ingrained in our cultural frames of reference that many of us cannot imagine eating meat that comes from the land around us rather than from a plastic container. Rather than recoil from this notion, we need to actively eradicate the toxic dualisms that inhibit us from engaging with the land in a healthy way. Of course, the current model of production and consumption will not persist—it cannot. The question is what will replace it. There are signs that people are already prepared to move beyond unmaking before it unmakes itself. Permaculture and modes of organic farming that nurture soil health are a step in the right direction. That people choose to pay more for organic food is promising too. Community gardens demonstrate that collectives can work together on shared land in a nonexploitive way.[22] In Australia, the Forestry Commission of New South Wales is working with Aboriginal communities to engage in what they call "cultural burning" to promote healthy forests. It's a step toward recognizing the active land-management practices of precontact Australians. And management of our most successful shared landscapes, our national parks, is increasingly being done in consultation with Aboriginal people. But at best these are small steps toward reconnection, at worst, tokenism within a dominant system of unmaking. To move forward as a species, we must eradicate the root metaphors of unmaking and replace them with the realities of connectedness. We need to reconfigure our worldviews to see the systems view of life and a world in which humans are deeply embedded.[23] Even the term "cultural burning" suggests a quaint Aboriginal practice that is somehow disconnected from heathy ecologies, rather than an active involvement in healthy landscapes. The dualisms of unmaking are false. Culture is nature, mind is body, soul is body, human is animal; there is no good and evil, only roles within ecosystems. Isolating ourselves and our domesticates might be working for the moment, but in the end it will be the undoing of us all. We need to embrace complexity and learn to trust land, plants, and animals in ways that allow life to flow across landscapes, sometimes toward and sometimes away from us. Only then will we be able to reconnect, not only with the deep webs of complex ecologies, but with our humanity.

After I remove the hive beetle larvae from the hive that I salvaged from Boggy Creek, I seal the gaps and wait to see if the bees will survive. Each morning I go to the hive and peer at the entrance to see if the bees are coming and going at a rate that might indicate that their numbers are increasing. I also put my ear to the hive entrance to listen for activity. It is buzzing in there, but not as loudly as my established hive, which also has far more activity at the entrance. I've chosen a spot in the garden under a sapote sapling, which provides shade only in the mornings. As the summer wears on, the days become warmer and the sun's glare reflects off the roof of the hive for too much of the day. I take an old bit of metal trellis, spike it into the ground next to the hive, and bend the top section over to create a frame. Then I take one of Leni's old baby swaddles and peg it to the trellis so the hive will be constantly shaded. I leave the hive alone to recover, but the activity within seems to be diminishing day by day. Four months after salvaging the hive, I remove the top section and the acetate viewing window to take a look. What I see is heartbreaking. The hive is silent and still but for three bees crawling over the labyrinth of propolis like lonely robots mechanistically fussing over the infrastructure of a lost civilization. Three days later, it is a ghost hive. No sign of any occupants, dead or alive. They've all flown off to fall dead upon the ground and turn to dust or be consumed by ants. Over my shoulder the waxing crescent moon maintains its smug silence; it has risen to my failure, to the folly of my effort to stave off the death of a hive. As I close the lid, I startle a skink, who darts for the cover of a nook in the retaining wall. I know, shortly after I'm gone, she will reemerge and resume her feeding under the warmth of the sun. A kookaburra laughs.

NOTES

FOREWORD

1. Steffen et al., "Anthropocene," 842.

2. Ripple et al., "World Scientists' Warning to Humanity"; Estrada et al., "Impending Extinction Crisis."

3. Bennett et al., "Broiler Chicken as a Signal," 1.

CHAPTER 1

1. Anthropologist Cris Shore details the ways in which neoliberalism has driven universities to what he refers to as "schizophrenia." The New Zealand government outlined thirteen different goals toward which universities should strive in order to earn foreign income, from education export, to Maori development, to acting as the critic and conscience of society. Essentially, this agenda jettisons the university's traditional roles in favor of a neoliberal ethos. See Shore, "Beyond the Multiversity."

2. Rose, "Slowly," 1.

3. See Fuentes, *Creative Spark*, for an extensive account of human niche construction—the ways in which humans shape the selective pressures that affect human evolution.

4. Rose, "Slowly," 2.

5. Alf Hornborg describes the ways in which the Mi'kmaq people turned modernity's logic against itself by co-opting its discourses in their project of resisting its forces. See his essay "Environmentalism, Ethnicity, and Sacred Places."

6. Leopold, *Sand County Almanac*, 132; Hornborg "Environmentalism, Ethnicity, and Sacred Places," 251.

7. Judt, *Ill Fares the Land*, 86.

8. See Worster's essay "Transformations of the Earth."

9. Lent, *Patterning Instinct*, 140.

10. Kim Sterelny and Trevor Watkins list what they consider "the cultural, cognitive and motivational" transformations associated with Neolithization, all of which are socioeconomic in nature. They do not delve into the conceptual shift that must occur when plants and animals are separated from their ecological contexts. "Neolithization in Southwest Asia," 675.

11. Rose, "Slowly," 2.

12. Ibid., 4.

13. I am not challenging the science of DNA testing here but rather the contextualization of the results.

14. Fuentes, *Creative Spark*, 23.

15. Lent, *Patterning Instinct*, 30.

16. Richard J. Williams and colleagues provide a comparison of empirical data from three freshwater habitats, two freshwater marine interfaces, two terrestrial habitats, and one island. Thus "two degrees of separation" in complex food webs is not a conclusive theory, but it does

have strong supporting evidence. See Williams et al., "Two Degrees of Separation."

17. Kortsch et al., "Climate Change Alters the Structure."

18. Monbiot, *How Did We Get into This Mess*, 16.

19. Judt, *Ill Fares the Land*, 38.

20. Peck and Tickell, "Neoliberalizing Space," 386.

21. Tsing, "Global Situation."

22. The Mont Pelerin Society actively seeks membership from non-Western countries, but not in order to enlighten its members; rather, its aim is to globalize the neoliberal project.

23. Plumwood, "Inequality, Ecojustice, and Ecological Rationality."

24. Humphrys and Cahill, "How Labour Made Neoliberalism."

25. Enterprise bargaining was later rebranded by the conservative Howard government with the suitably Orwellian term "work choices." It proved politically disastrous.

26. The RIRDC has since been rebranded as AgriFutures. Despite the rebranding, it performs the same role.

27. RIRDC, "Annual Report 2007–2008."

28. Franklin, *Animal Nation*.

29. Ingold, *Perception of the Environment*, 47. For Diamond's argument, see his article "Worst Mistake."

30. Pearce-Duvet, "Origin of Human Pathogens." Pearce-Duvet suggests that domesticated animals may have provided a "stable conduit" by which humans were infected by wildlife diseases. I think that imposing sedentism, crowding, and human interventions on these animals would also have made for disease hubs by which humans were affected.

31. Larsen, "Agricultural Revolution as Environmental Catastrophe."

32. Anthropologist Bruce Smith proposes a cultural niche construction theory for domestication that holds that cultivating plants and keeping animals created a novel niche to which domesticates and humans have adapted. Smith, "Cultural Niche Construction Theory."

33. In saying this, I recognize that there probably were imbalances in power relations among hunter-gatherers, under which young girls, especially, were less empowered. But these power imbalances were not enforceable by directly withholding resources from the less empowered. Rather, they were rooted in myth and tradition.

34. Leach, "Human Domestication Reconsidered."

35. Perry et al., "Diet and the Evolution of Human Amylase."

36. Galvani and Slatkin, "Evaluating Plague and Smallpox."

37. Greger Larson and Dorian Fuller list some of the changes that occurred in domesticated animals: endocrine changes, increased docility, altered reproduction pattern and output, altered coat color, floppy ears, facial neoteny, usually a reduction in size, and other changes in body proportions. See Larson and Fuller, "Evolution of Animal Domestication."

38. See Guintard, "On the Size of the Ure-ox." Note, though, that some have questioned this height and suggest that it's closer to 150–80 cm (59–71 in.).

39. Some have argued that early domestic dogs subsisted on human feces. Whether a band of hunter-gatherers produces enough poo to sustain a reproductively viable population of dogs is open to question. See Sarah Marshall-Pescini et al., "Integrating Social Ecology."

40. Mietje Germonpré and colleagues present evidence of a prehistoric dog dating to 31700 B.P., although the status of this individual based on DFA (discriminant function analysis) has been challenged by Darcy Morey, who suggests that three of the five specimens in the reference group used should probably be classified as wolves. Mikhail Sablin and Gennady Khlopachev found the remains of what

appear to be dogs in deposits at Eliseevichi in central Russia. There, two crania of adult dogs were found and dated to 13000–17000 B.P. Interestingly, a hole had been made in one of the crania so that the brain could be extracted. From Saint-Thibaud in France, Louis Chaix describes a small dog in the deposits of a rock shelter. The remains were dated to 10050 (plus or minus a hundred years), and the tooth wear and unfused trochanter indicate an individual of about ten months old. These all predate settled agriculture in their respective locales. See Germonpré et al., "Fossil Dogs and Wolves"; Morey, *Dogs*, 21; Sablin and Khlopachev, "Earliest Ice Age Dogs"; and Chaix, "Preboreal Dog from the Northern Alps."

41. One problem with Pierotti and Fogg's hypothesis is that they base it on accounts of North American wolves, who exhibit markedly different ways of interacting with humans than European wolves, who have a history of predation on humans. See Kruuk, *Hunter and Hunted*, 69–73.

42. Zeder, "Archaeological Approaches to Documenting Domestication."

43. Zeder lists three pathways to domestication: commensal, prey, and directed. See Zeder, "Domestication of Animals."

44. Bradley and Magee, "Genetics and Origins of Domestic Cattle."

45. Olsen, "Early Horse Domestication."

46. For a comprehensive list of early domesticates, see Larson and Fuller, "Evolution of Animal Domestication."

47. Zeder, "Pathways to Animal Domestication," 250.

48. Clutton-Brock, introduction to *Walking Larder*, 7.

49. Fuentes, "Monkey and Human Interconnections," 128.

50. See Baynes-Rock and Teressa, "Shared Identity of Horses and Men."

51. David Orton argues that animals owned by humans constitute "sentient property" in that they have agency. This means that a wild/domestic distinction needs to be based more on how animals "play an active role in human society." See Orton, "Both Subject and Object," 188.

52. Zeder, "Domestication of Animals," 163–64.

53. Fijn, *Living with Herds*, 19, 35, 201.

54. Orton, "Both Subject and Object," 193.

55. Larson and Fuller, "Evolution of Animal Domestication," 124.

56. There seems to be more than a one-way selective process involved here, as the foxes would almost surely be responding to the body language of the farmworkers. I don't think this factor undermines the experiment, but it was not controlled for.

57. See Wilkins, Wrangham, and Fitch, "'Domestication Syndrome' in Mammals." These authors point to deficits in neural crest cell development during embryonic development, and their limiting function over the adrenal glands, as key to domestication syndrome. They differ from Trut and colleagues in arguing that these changes involve multiple "loss-of-function" mutations in the genes involved.

58. Trut, Oskina, and Kharlamova, "Animal Evolution During Domestication," 356.

59. Shepard, *Tender Carnivore*, 10.

60. Jean-Denis Vigne notes that before morphological changes can take place through domestication, there must be some change in the animals' ways of life, such as a reduction in their mobility, altered diet, or a move to a different environment. See Vigne, "Early Domestication and Farming," 138. Melinda Zeder argues similarly that evidence of domestication can be present in the archaeological record one thousand years prior to morphological changes in the target species. Zeder, "View from the Zagros," 139.

61. Leach, "Selection and the Unforeseen Consequences."

62. Vigne et al., "Early Process of Mammal Domestication."

63. Fuentes and Kohn, "Two Proposals."

64. Fuentes, *Creative Spark*, 121.

65. Eve Vincent and Timothy Neale criticize what they call an environmentalist narcissism that expects "Aboriginal people to play the part of the cultural other to Western capitalist modernity." This criticism is not without foundation, but it does not negate the fact that there is a crucial difference between Aboriginal and colonial Australian ways of engaging with the land that is evidenced in the changes to the landscape and its species that have occurred over the past 230 years. See Vincent and Neale, "Unstable Relations," 307.

66. See Rose, *Dingo Makes Us Human*.

67. Trigger, "Indigeneity, Ferality, and What 'Belongs.'"

68. Rose, "Indigenous Philosophical Ecology," 303.

69. See Wheeler, *Intertribal Relations in Australia*.

70. Lien, *Becoming Salmon*, 20.

71. At the same time, I was undertaking another project, surveying and interviewing farmers who were having problems with wild dogs.

CHAPTER 2

1. The origin of the first dingoes to enter Australia has not been established beyond doubt. The story of dingoes coming with hunter-gatherer people from Sulawesi is based on genetic evidence from humans and dingoes and archaeological evidence of links between Borneo, Sulawesi, and Australia. See Fillios and Taçon, "Who Let the Dogs In?"

2. On the basis on recent ethnographic and archaeological material, Margaret Titcomb and Mary Kawena Pukui argue that the early seafarers of the Pacific brought dogs along on their voyages for multiple purposes. Not only did dogs provide meat and skins for clothing; they held great religious significance, and their teeth, bones, and hair were used in various rituals. Titcomb and Pukui, *Dog and Man in the Ancient Pacific*.

3. Bradley Smith and Peter Savolainen explain how dingoes represent a genetic bottleneck—the founder population carrying only one mtDNA haplotype and two Y-chromosome haplotypes. See Smith and Savolainen, "Origin and Ancestry of the Dingo," 69. Kylie Cairns and Alan Wilton argue for at least two immigration events or a much earlier single immigration event after which two separate lineages evolved, one focused in the northwest and one in the southeast. See Cairns and Wilton, "History of Canids in Oceania."

4. Mike Letnic and colleagues suggest that dingoes were not just competitors with thylacines but may well have been killing them. This theory is based on size comparisons and models of optimal prey size for dingoes. Thylacines and Tasmanian devils persisted in Tasmania not least because dingoes never crossed over to the island. Thylacines became extinct in Tasmania in the twentieth century as a result of eradication efforts by Europeans; the last thylacine died at the Beaumaris Zoo in Hobart in 1936. See Letnic, Fillios, and Crowther, "Extinction of the Thylacine."

5. Francis, *Domesticated*, 4.

6. Breckwoldt, *Very Elegant Animal*, 63.

7. Tindale cited in ibid., 65. Laurie Corbett says that the Garawa of the Gulf of Carpentaria ate dingoes all year round, although they weren't a staple. See Corbett, *Dingo in Australia*, 20.

8. See Young, "Dingo Scalping and the Frontier Economy."

9. Rose, *Dingo Makes Us Human*, 176.

10. Fijn, "Dog Ears and Tails."

11. Hamilton, "Aboriginal Man's Best Friend," 294. Roland Breckwoldt raises the

question of whether dingoes were used for keeping warm at night. This practice has indeed been documented among Aboriginal people, and it led to the Australian colloquialism used to describe a night that is very cold: a "three dog night." But as Breckwoldt points out, canine bedfellows tend to seek out the best position for themselves—between campfire and warm human. This is a perfect position in the short term, but it causes a lot of irritation to the humans who are cut off from the fire and have the cold night air at their backs. See Breckwoldt, *Very Elegant Animal*, 64.

12. Hamilton, "Aboriginal Man's Best Friend," 288.

13. Ibid., 294.

14. Breckwoldt, *Very Elegant Animal*, 60.

15. This observation comes from Meggitt's work among the Walbiri. "Australian Aborigines and Dingoes," 11.

16. In *First Domestication*, Pierotti and Fogg present an argument that wolves and humans became close simply by being amicable neighbors.

17. Rose, *Dingo Makes Us Human*, 105.

18. Berndt and Berndt, *Speaking Land*, 29–32.

19. Breckwoldt, *Very Elegant Animal*, 91.

20. Quoted in Main, *Gunderbooka*, 33.

21. Young, "Dingo Scalping and the Frontier Economy," 101.

22. This is not to say that dingoes were eradicated within the fenced-off part of the country or that farmers ceased to have problems with them. The dingo fence merely prevented incursions from the arid lands, while inside the boundary, resident dingoes were killed.

23. Parker, "Bringing the Dingo Home," 50.

24. Smith, "Howl and the Pussy."

25. Parker, "Bringing the Dingo Home," 282.

26. Another attack occurred on K'gari in April 2019, when a dingo dragged a fourteen-month-old boy from a camper. Australian Broadcasting Corporation, "Dad Fights off Dingoes."

27. Parker, "Bringing the Dingo Home," 223.

28. Corbett, *Dingo in Australia*, 165.

29. Pierotti and Fogg show how hybridity is characteristic of canids, many species of which can and do interbreed. See *First Domestication*, 42.

30. Probyn-Rapsey, "Dingoes and Dog-Whistling."

31. According to Gwen, the samples undergo twenty tests for fourteen genetic markers that distinguish dingoes from domestic dogs.

32. A Yowie is a hominid creature of Aboriginal folklore. It is akin to a Sasquatch or Yeti, having a hairy body and big feet.

33. A recent study found that the presence of dingoes affects herbivore populations, which affect plant communities, which affect the size and shape of sand dunes, which affect the wind. The study linked trophic cascades to changes in desert dune geomorphology using high-resolution drone data. See Lyons et al., "Changes in Desert Dune Geomorphology."

CHAPTER 3

1. See Crittenden, "Importance of Honey Consumption."

2. Wrangham, "Honey and Fire in Human Evolution," 156. Serendipitously, Wrangham's article was published the same year as Crittenden's.

3. See Boesch, Head, and Robbins, "Complex Tool Sets for Honey Extraction."

4. See Crane, *World History of Beekeeping*, 330.

5. See Kraft and Venkataraman, "Could Plant Extracts Have Enabled Hominins?"

6. Ancient Egypt does fit the bill as regards the environment in which

beekeeping might emerge: plenty of forage, sedentism, and a demand for honey and wax, all of which would result from an increasing population and specialized labor. The same factors fostered beehive keeping in Mayan civilization.

7. See Guy, "Honey Hunters of Southern Africa."

8. Crane, *World History of Beekeeping*, 109 (thorns), 147 (Malayan sun bears), 132 (brown bears in Poland).

9. Ibid., 127–29. These practices have since disappeared, along with the forests that harbored them.

10. For a detailed description of Ethiopian beekeeping practices, see Abebe, "Indigenous Knowledge of Beekeeping Practices."

11. Underhill's blog documents the processes of keeping honeybees and the various threats to beekeeping. Richard Underhill, "Ethiopia's Limich," *Peace Bee Farmer*, March 15, 2012, http:// peacebeefarm.blogspot.com.au/2012/03/ethiopias-limich.html.

12. Crane, *World History of Beekeeping*, 405–6.

13. Ibid., 423.

14. See Delaney et al., "Genetic Characterization of Commercial Honey Bee."

15. During their honeymoon, which they spent searching for native bees' nests in northwestern Australia, Anne and Les Dollin were told that Aboriginal people used to harvest nests and seal them up with mud so that they could recover and be harvested again in the future. They didn't provide any evidence of this, but that doesn't mean it didn't occur. Dollin and Dollin, "Tracing Aboriginal Apiculture."

16. So it is not just archaeologists who see phalluses everywhere. See McKnight, "Sexual Symbolism of Food."

17. Akerman, "Honey in the Life of the Aboriginals," 176.

18. Welch, "Beeswax Rock Art."

19. Fijn's book *Living with Herds* documents a very early example of multispecies ethnography in which livestock animals are considered social actors. I should also note that Debbie Rose was one of Natasha's supervisors and has influenced her ideas and writing.

20. See Fijn's article "Sugarbag Dreaming," and the accompanying video, uploaded March 11, 2014, at https://vimeo.com/88737231.

21. Fijn and Baynes-Rock, "Social Ecology of Stingless Bees."

22. Megan Halcroft and colleagues found that these programs are becoming widespread, from the Wik-Mungkan in North Queensland to the Gumbaynggirr people of northern New South Wales. See Halcroft, Spooner-Hart, and Dollin, "Australian Stingless Bees."

23. Letter to the editor, *Aussie Bee*, November 1997, 21.

24. Between 1998 and 2010, the number of hives of stingless bees in Australia rose from 1,429 to 4,935. Halcroft et al., "Australian Stingless Bee Industry."

25. Heard, *Australian Native Bee Book*, 34.

26. Ibid., 182.

27. Freya Mathews has written a thoughtful article about what exactly we find so disturbing about colony collapse disorder. She suggests that it is the threat to ecological systems that lies at the heart of the ethical concerns about CCD. Mathews, "Planet Beehive."

28. Massaro et al., "Anti-Staphylococcal Activity."

29. Mark Greco and colleagues call this the "alternative pharaoh approach." When bees notice a beetle in the hive, they mount a continuous attack, which causes the beetle to adopt a "turtle" defense. This stops the beetle from moving about and gives the bees a chance to daub the invader with resin and mud, mummifying the beetle alive. It's an approach similar to that of honeybees, who "encapsulate"

invaders, but is in fact a case of convergent evolution rather than shared behavior. Honeybees confine invaders to cells rather than coat them in resin. See Greco et al., "Alternative Pharaoh Approach."

CHAPTER 4

1. This is not even the most expensive of their products. At the time of this writing, the same company offers a crocodile skin backpack for US$81,500.

2. Quammen, *Monster of God*, 163. Freshwater crocodiles were already protected by then.

3. Queensland was also the last state to remove voting restrictions for Aboriginal people.

4. An outspoken Queensland member of Parliament recently called for a return to crocodile culling on the basis of the danger they pose to humans, but his claim that a crocodile kills someone every three months was shown to be false. The truth is that crocodiles cause only one human fatality roughly every three years in Australia. Nationally, this is less than one-third the number of deaths caused by dogs. Australian Broadcasting Corporation, "Fact Check"; National Coroners Information System, "Animal-Related Deaths."

5. Tisdell and Nantha, "Farming of Saltwater Crocodiles."

6. Quammen, *Monster of God*, 168.

7. Jan van der Ploeg and colleagues challenge the assumption that people need economic incentives to conserve crocs. In a study in the Philippines they found that local pride and appreciation of the rarity of crocs also mobilized the community to undertake conservation efforts. Ploeg, Araño, and Weerd, "What Local People Think About Crocodiles."

8. Jon Hutton and Grahame Webb, however, suggest that the benefits of a closed farm in terms of conservation may be "minimal or non-existent." Hutton and Webb, "Principles of Farming Crocodiles."

9. While sustainable use might work for crocodiles, for something like a rare fish species, the concept invites disaster. If potentially anyone has access to the fish, then there is no limit to how many will be taken. Of course, allowing trade in the species might foster captive breeding programs, as with the Asian arowana, but this is not a conservation model; it is an economic one. If the fish are no longer integrated within the complex ecologies of their habitats, then they are but shadows of fish.

10. This is not necessarily always the case. In World Heritage Site Chiribiquete National Park, Colombia, the Indigenous people are recognized, and interference with their traditional occupation and use of the land is prohibited.

11. Hornborg, "Environmentalism, Ethnicity, and Sacred Places," 258.

12. Plumwood, "Being Prey."

13. Fijn, "Living with Crocodiles," 15.

14. Marcus Barber relates a story in which barramundi fishermen strung up a crocodile head at a place that the Yolngu consider the home of the ancestral crocodile, named Baru. This was an immensely provocative act, and not just because crocodile heads are held to be sacred. It occurred just before the native title claim hearings of 2004, in which the Yolngu claimed their rights to traditional lands under the Native Title Act of 1993. Barber, "Where the Clouds Stand," 9.

15. Morphy, "Reproduction of the Ancestral Past," 198.

16. Berndt and Berndt, *Speaking Land*, 313.

17. Ibid., xxv.

18. Gagadju is a language group of northwestern Arnhem Land.

19. For another such novel camping party trick, see chapter 6 on kangaroos.

20. Berndt and Berndt, *Speaking Land*, 220.

21. Ibid., 252.
22. Ibid., 321–22.
23. Ibid., 426.
24. Gordon, *Milbi*, 9–11. For a long time, Gordon was forced to conceal his knowledge of Aboriginal lore because of prohibitions by the church. He was also an accomplished artist.
25. Berndt and Berndt, *Speaking Land*, 400.
26. Heath, *Nunggubuyu Myths*, 235–39.
27. Halfway through the Nunggubuyu story collected by Heath, the storyteller changes the noun class for the crocodile from nonhuman to human and then back again.
28. In "Principles of Farming Crocodiles," Hutton and Webb promote routine and repetition in crocodile farming to the extent that workers "wear the same colour overalls, enter the enclosures in the same place and clean in exactly the same way on every occasion" (24).
29. Farming cassowaries proved to be unsuccessful. The males in the wild required large home ranges, and without the right variety of fruits and berries, they failed to produce sperm.
30. Interestingly, crocodiles are very probably descended from endothermic ancestors. Physiological, anatomical, and developmental features of the crocodilian heart align with paleontological evidence that the ancestors of living crocodilians were active and endothermic. The lineage reverted to ectothermy when it came to occupy the aquatic ambush-predator niche. I shudder to think what a crocodilian might be like if it were a highly active warm-blooded predator. Seymour et al., "Evidence for Endothermic Ancestors."
31. The sex of crocodiles is determined by the temperature of the egg at incubation. Eggs kept at 32–33 degrees Celsius (89.6 to 91.4 degrees Fahrenheit) result in male hatchlings; higher or lower temperatures result in females. See also Hossain et al., "High Hatching Success."
32. Nile crocodiles in Botswana are also susceptible to fungal infections. See Dzoma, Sejoe, and Segwagwe, "Commercial Crocodile Farming in Botswana."
33. While in Papua New Guinea, Mike developed a method for humane slaughter that is currently used at Bindara. A chock is placed under the crocodile's neck and his head is bent forward. This compresses the spine against the skin and opens up the vertebrae. A surgical knife is used to make a cut across the back of the neck and directly through the spinal column, whereupon death should be instantaneous. This is followed by a process called "pithing," in which a metal rod is inserted into the spinal column to shatter the brain, just to make certain that the crocodile is dead.
34. It should be noted that a crocodile's penis looks even more terrifying than a crocodile.
35. Johnston et al., "Semen Collection and Seminal Characteristics."
36. Sorensen, "Scientists Using Artificial Insemination."
37. Johnston et al., "Semen Collection and Seminal Characteristics," 33–34.
38. Hornborg, "Environmentalism, Ethnicity, and Sacred Places," 258.

CHAPTER 5

1. The Kamilaroi story of Emu and Brolga is a widespread myth that explains why emus are flightless. After Emu convinces Brolga to kill all but two of her chicks, Brolga convinces Emu to burn off her wings. Austin and Tindale, "Emu and Brolga."
2. Tuinen, Sibley, and Hedges, "Biogeography of Ratite Birds."
3. John Harshman and colleagues accept the more parsimonious explanation that tinamous might have independently regained the ability to fly, but the regularity with which bird species have lost the ability to fly does lend credence to independent loss of flight in the three other

ratites. Harshman et al., "Phylogenomic Evidence."

4. Maddock, "Emu Anomaly."

5. There's a tragic inevitability in the stories of Emu losing the ability to fly. But is it loss or adaptation? We don't think of human bipedality as a loss of tree-climbing ability.

6. Being unable to find a collective noun for sea monsters, I've settled for one that applies to sea lions.

7. Davies, "Food of Emus."

8. There's an ongoing debate about what name these people used for themselves and what language they spoke. I use Guringai here, but Val Attenbrow argues that they were Darug-speaking people. See Attenbrow, *Sydney's Aboriginal Past.*

9. Norris and Hamacher, "Astronomy of Aboriginal Australia."

10. Details of the interpretations of the sky emu differ across groups. Norris and Norris, *Emu Dreaming*, 5.

11. Fuller et al., "Emu Sky Knowledge," 2.

12. Mathews, *Opal That Turned into Fire*, 107.

13. Interestingly, the builders of bora rings for these ceremonies aligned them with the Milky Way. See Fuller et al., "Emu Sky Knowledge," 7.

14. Ibid.

15. Garde, *Something About Emus*, xviii.

16. Emus eat fruits, flowers, seeds, grass shoots, herbs, shrubs, and insects. Davies, "Natural History of the Emu," 111.

17. Garde, *Something About Emus*, 69.

18. Michael Pickering mentions that the Garawa hunted emus by waving a cloth or object, or even lying in the grass and waving a foot in the air. Pickering, "Garawa Methods of Game Hunting," 16.

19. Garde, *Something About Emus*, 34.

20. Ibid., 10.

21. For the Nunggubuyu of southeast Arnhem Land, emus are similarly connected to stingless bees, as they are a submoiety of the Yirritja moiety, which is complementary to the Dhuwa moiety. Yirritja is a kind of stingless bee (their local names are Mandayung [Dhuwa] and Mandaridja [Yirritja]). Leeden, "Thundering Gecko and Emu."

22. Debbie Rose describes a man who shot an emu, whereupon some children present raced to club it to death. The group returned to camp and gave the carcass to a person of the emu clan. People of the emu clan are prohibited from eating emu meat but are responsible for the care of dead emus and distribution of the meat. Rose, *Dingo Makes Us Human*, 84.

23. Garde, *Something About Emus*, 81.

24. The uniform morphology of emus in Australia suggests that the whole population moves as freely as the other birds in Western Australia do. Davies, "Natural History of the Emu," 112.

25. Garde, *Something About Emus*, 38.

26. Davies, Beck, and Kruiskamp, "Results of Banding 154 Emus."

27. Robin, "Emu: National Symbols," 246.

28. See Johnson, "'Feathered Foes.'"

29. Ibid., 153.

30. Emus are still considered a pest, however, and baiting programs are used to try to control their numbers. Australian Government, "Emu Control."

31. Nicholls, "Commercial Emu Raising."

32. O'Malley and Snowden, "Emu Products."

33. Ibid., 4.

34. Sales, "Emu (*Dromaius novaehollandiae*)," 14.

35. Emus don't grow and mature as rapidly as chickens do. Ibid., 6.

36. Ever since the development of the commercial emu industry began around 1990, most of the attention has been focused on oil production. Ibid., 8.

37. A hundred-milliliter bottle of emu oil (about 3.38 ounces) retails for about US$25.00.

38. Peter Ghosh, Michael Whitehouse, Michael Dawson, and Athol G. Turner,

"Anti-Inflammatory Composition Derived from Emu Oil," U.S. Patent 5,431,924, July 11, 1995, https://patents.justia.com/patent/5431924.

39. These are the results of a study at the University of Queensland using laboratory-bred mice to test anti-inflammatory effects of emu oil.

40. Jeengar et al., "Review on Emu Products."

41. Males protect chicks and attack threats, including the chicks' biological mothers. Davies, "Natural History of the Emu," 113.

42. The amino acid requirements of growing emus have been estimated at age intervals with the use of a mathematical model used for commercial poultry production. O'Malley and Snowden, "Emu Products"; Sales, "Emu (*Dromaius novaehollandiae*)."

43. Malecki and Martin, "Emu Farming"; Sales, "Emu (*Dromaius novaehollandiae*)," 4.

44. Warale et al., "Emu Farming," 10. These authors recommend separating males and females after breeding.

45. Emus are susceptible to diseases like encephalomyelitis, clostridiosis, salmonellosis, aspergillosis, etc. Internal and external parasitic infestations like coccidiosis, ascaridiosis, and lice are not uncommon.

46. Research into the effects of declawing has been undertaken by the South Australian Research and Development Institute.

47. No less than seven avian influenza subtypes have been isolated from emus in the United States and China. Sales, "Emu (*Dromaius novaehollandiae*)," 13.

48. Emu muscles have even been assigned trade names. Ibid., 10.

CHAPTER 6

1. Osborne, "Acclimatizing the World," 135.

2. Singley, "'Fit for Man to Eat,'" 37.

3. Boom and Ben-Ami, "Shooting Our Wildlife," 34. Tasmanian brushtail possums and crocodiles are two other land-based native species that are harvested commercially.

4. Boom et al., "'Pest' and Resource," 21; Singley, "'Fit for Man to Eat,'" 28.

5. Croft, "Kangaroos Maligned," 28.

6. This is not to say that nobody ate kangaroo at that time. Rural dwellers continued to eat kangaroo throughout the twentieth century; the first Australian cookbook (*The English and Australian Cookery Book*, published in 1864) had a dozen recipes for kangaroo dishes.

7. John Auty argues that kangaroo numbers have in fact diminished since colonization. He bases this argument on historical accounts that mention abundant kangaroos and the inability of Aboriginals to use dingoes to hunt them (which Auty takes to mean that their numbers weren't checked by people or dingoes). Auty, "Red Plague, Grey Plague."

8. In 2017, the total population was estimated at 47,226,027. The killing quota was 7,174,072, and the total harvest was 1,632,098. In New South Wales, Queensland, and South Australia, males constituted between 94 and 99.9 percent of the total harvest (1,419,492 individuals). In Western Australia, males constituted between 80.3 and 81.4 percent of the total harvest (68,777 individuals). Australian Government, "Macropod Quotas and Harvest."

9. The CSIRO also kept captive pademelons in "self-perpetuating colonies." Calaby and Poole, "Keeping Kangaroos in Captivity," 5.

10. Shepherd, "Feasibility of Farming Kangaroos," 43.

11. Mitchell, *Journal of an Expedition*, 320.

12. Boom et al., "'Pest' and Resource," 21; Pascoe, *Dark Emu, Black Seeds*, 42–43.

13. Murphy and Bowman, "Interdependence of Fire."

14. Bird et al., "'Fire Stick Farming' Hypothesis."

15. Gammage, *Biggest Estate on Earth*, 6, 7, 8, 9, 14, 15, 16, 39, 41.

16. Ronald Berndt heard this story in Djambarbingu in 1949 from a man named Danggoubwi. Berndt and Berndt, *Speaking Land*, 178. On the dearth of kangaroo protagonists in Aboriginal stories, see also Mathews, *Opal That Turned into Fire*.

17. Ian Abbot argues that Aboriginal people exterminated kangaroos and wallabies from Australia's smaller islands (those smaller than ninety hectares), but only where they had watercraft capable of making a crossing. Abbott, "Aboriginal Man as an Exterminator."

18. See Altman, "Dietary Utilisation of Flora and Fauna." Altman spent a year in the field with the Gunwinggu of north-central Arnhem Land—neighbors of the Yolngu. He found, among other things, that the seasons, movements of people, rituals, and food availability were all intimately connected.

19. See Pickering, "Garawa Methods of Game Hunting."

20. David Croft discusses the ethics of incapacitating kangaroos to be eaten later. "Kangaroos Maligned," 21.

21. Cawte, Djagamara, and Barrett, "Meaning of Subincision," 246.

22. Ibid., 249.

23. Quoted in Main, *Gunderbooka*, 22.

24. Mathews, *Opal That Turned into Fire*, 8.

25. New South Wales Government, "Brewarrina Aboriginal Fish Traps."

26. Quoted in Main, *Gunderbooka*, 17 (Sturt), 27 (Randell). It's possible that much of this was hyperbole aimed at encouraging settlers to come to the district. But given that these same explorers were quick to highlight the problems with the "blacks" in the district, it's not a given that they were exaggerating the qualities of the country.

27. Some see woody weed as the formation of scar tissue as a result of environmental damage. Ibid., 63.

28. Ibid., 30.

29. A road train is a trucking vehicle used in remote areas, consisting of two or more trailers or semitrailers hauled by a prime mover.

30. A 2003 study by Margaret Chapman found a marked difference in the ways farmers perceived goats and kangaroos: 81 percent saw kangaroos as more difficult to control than goats. Only 4 percent saw goats as more difficult to control. Chapman, "Kangaroos and Feral Goats," 26.

31. See Plumwood, "Animals and Ecology."

32. Lawns in the region stretch uninterrupted from stoop to street; the local council has historically discouraged sidewalks lest they encourage economically destructive practices such as walking and cycling.

33. Maria Ignatieva and Marcus Hedblom provide a critique of modern lawns and a viable alternative in "Alternative Urban Green Carpet."

34. Zeder, "Pathways to Animal Domestication," 231.

35. State and territory legislation holds that wild animals are protected, so licenses are required to kill them. Boom and Ben-Ami, "Shooting Our Wildlife," 18.

36. McCullough and McCullough, *Kangaroos in Outback Australia*, 112.

37. Ibid., 119.

38. Priddel, Shepherd, and Ellis, "Homing by the Red Kangaroo."

39. Viggers and Hearn, "Kangaroo Conundrum." A similar study also needs to be done in the rangelands.

40. McCullough and McCullough, *Kangaroos in Outback Australia*, 98–99.

41. Moree is actually 445 kilometers by road from Bourke, but in the context of western New South Wales, it is nearby.

CHAPTER 7

1. In a study using MRI scans of rabbit brains, researchers found that in comparison to wild rabbits, domesticated rabbits exhibited a reduction in size of the amygdala and enlargement in the prefrontal cortex. The amygdala is associated with the fear response, and the prefrontal cortex is associated with modulation of negative affect. In other words, the study shows how tameness is not necessarily intrinsic to a wild species but can be enhanced in a species undergoing domestication.

2. Greger Larson and Dorian Fuller argue that people could not have begun intentionally domesticating animals "until they had procured them through entirely unintentional means." "Evolution of Animal Domestication," 127.

3. Leach, "Selection and the Unforeseen Consequences," 71.

4. Kylie Cairns and Alan Wilton argue for two subpopulations based on lineages from two immigration events. "History of Canids in Oceania."

5. The dates given here are the subject of debate, as they are based on the comparisons of bone sizes in archaeological assemblages and in some cases on small sample sizes. West and Zhou, "Did Chickens Go North?"

6. Milder alternatives to antibiotics, known as ionophores, are being used in some regions to control gut microbiota. These have no impact on human health in terms of antimicrobial resistance. But they are being phased out in favor of more expensive probiotic feed additives and vaccines.

7. Shepard points out that the response to chemical-resistant pests—the replacement of one chemical control with another—is "consumerism at its best." *Tender Carnivore*, 251.

8. Where I live in New South Wales, the local council has a special classification for land previously used for banana farming because of the legacy of chemical residues in the soil.

9. Williams, "Inhumanity of the Animal People."

10. Fijn, *Living with Herds*, 208.

11. Plumwood, "Animals and Ecology," 4.

12. Hall, "Beyond the Human," 380.

13. Shepard, *Tender Carnivore*, 266.

14. Rose, James, and Watson, *Indigenous Kinship with the Natural World*, 1.

15. Rose "Indigenous Philosophical Ecology," 296.

16. Quandongs are a traditional bush fruit that can be used in jams and pies. Development of commercial cultivars began in 1973, and by the 1990s they were being harvested. But the orchards were severely affected by drought and moths. The Australian Quandong Industry Association, formed in 1992, is no longer operational, but there are still some twenty-five commercial growers in Australia. AgriFutures Australia, "Quandong."

17. Rose, James, and Watson, *Indigenous Kinship with the Natural World*, 19.

18. This painting appears in an open access article. See Bird et al., "Size of Hunter-Gatherer Groups," 102.

19. Tony Judt presents a series of charts that demonstrate correlations between inequality and all manner of social ills. Judt, *Ill Fares the Land*, 29–33.

20. Monbiot, *Feral*, 153. See also Harvey et al., "Rise of Mega Farms."

21. Monbiot, *Feral*, 84.

22. See Samiei, "Architecture and Urban Ecosystems."

23. Jeremy Lent regards this as "the great transformation." *Patterning Instinct*, 434.

BIBLIOGRAPHY

Abbott, Ian. "Aboriginal Man as an Exterminator of Wallaby and Kangaroo Populations on Islands Round Australia." *Oecologia* 44, no. 3 (1979): 347–54.

Abebe, Workneh. "Identification and Documentation of Indigenous Knowledge of Beekeeping Practices in Selected Districts of Ethiopia." *Journal of Agricultural Extension and Rural Development* 3, no. 5 (2011): 82–87.

AgriFutures Australia. "Quandong." May 24, 2017. https://www.agrifutures.com.au/farm-diversity/quandong.

Akerman, Kim. "Honey in the Life of the Aboriginals of the Kimberleys." *Oceania* 49, no. 3 (1979): 169–78.

Altman, Jon C. "The Dietary Utilisation of Flora and Fauna by Contemporary Hunter-Gatherers at Momega Outstation, North-Central Arnhem Land." *Australian Aboriginal Studies* 1 (1984): 35–46.

Attenbrow, Val. *Sydney's Aboriginal Past: Investigating the Archaeological and Historical Records*. Sydney: University of New South Wales Press, 2010.

Austin, Peter, and Norman B. Tindale. "Emu and Brolga, a Kamilaroi Myth." *Aboriginal History* 9, no. 1 (1985): 8–21.

Australian Broadcasting Corporation. "Dad Fights off Dingoes That Took Toddler from Camper Trailer on Fraser Island off Queensland Coast." April 20, 2019. https://www.abc.net.au/news/2019-04-19/queensland-toddler-attacked-by-dingo-on-fraser-island/11031828.

———. "Fact Check: Is a Person Torn to Pieces by a Crocodile Every Three Months in North Queensland?" July 12, 2018. http://www.abc.net.au/news/2017-11-30/fact-check-does-a-crocodile-kill-someone-every-three-months-/9202902.

Australian Government, Department of Primary Industries and Regional Development. "Emu Control: Strychnine Landholder Information." March 9, 2015. https://www.agric.wa.gov.au/strychnine/emu-control-strychnine-landholder-information-0.

Australian Government, Department of the Environment and Energy. "Macropod Quotas and Harvest for Commercial Harvest Areas in NSW, QLD, SA and WA, 2016–19." http://www.environment.gov.au/system/files/pages/ee20f301-6c6c-44e4-aa24-62a32d412de5/files/kangaroo-statistics-states-2018.pdf.

Auty, John. "Red Plague, Grey Plague: The Kangaroo Myths and Legends." *Australian Mammalogy* 26, no. 1 (2004): 33–36. https://doi.org/10.1071/AM04033.

Barber, Marcus. "Where the Clouds Stand: Australian Aboriginal Attachments to Water, Place, and the Marine Environment in Northeast Arnhem Land." PhD diss., Australian National University, 2005.

Baynes-Rock, Marcus, and Tigist Teressa. "Shared Identity of Horses and Men in Oromia, Ethiopia." *Society and Animals*. Forthcoming.

Beach, Eric. "Emus Out of Genoa." In *Weeping for Lost Babylon*. Sydney: Angus and Robertson in association with Paper Bark Press, 1996.

Bennett, Carys E., Richard Thomas, Mark Williams, Jan Zalasiewicz, Matt Edgeworth, Holly Miller, Ben Coles, et al. "The Broiler Chicken as a Signal of a Human Reconfigured Biosphere." *Royal Society Open Science* 5, no. 12 (2018): 1–11. http://doi.org/10.1098/rsos.180325.

Berndt, Ronald M., and Catherine H. Berndt. *The Speaking Land: Myth and Story in Aboriginal Australia*. Rochester, Vt.: Inner Traditions, 1994.

Bird, Douglas W., Rebecca Bliege Bird, Brian F. Codding, and David W. Zeanah. "Variability in the Organization and Size of Hunter-Gatherer Groups: Foragers Do Not Live in Small-Scale Societies." *Journal of Human Evolution* 131 (2019): 96–108.

Bliege Bird, Rebecca, Douglas W. Bird, Brian F. Codding, Christopher H. Parker, and James H. Jones. "The 'Fire Stick Farming' Hypothesis: Australian Aboriginal Foraging Strategies, Biodiversity, and Anthropogenic Fire Mosaics." *Proceedings of the National Academy of Sciences* 105, no. 39 (2008): 14796–801. https://doi.org/10.1073/pnas.0804757105.

Boesch, C., J. Head, and M. M. Robbins. "Complex Tool Sets for Honey Extraction Among Chimpanzees in Loango National Park, Gabon." *Journal of Human Evolution* 56, no. 6 (2009): 560–69.

Boom, Keely, and Dror Ben-Ami. "Shooting Our Wildlife: An Analysis of the Law and Policy Governing the Killing of Kangaroos." Report by THINKK, the Kangaroo Think Tank at the University of Technology Sydney, 2010.

Boom, Keely, Dror Ben-Ami, David B. Croft, Nancy Cushing, Daniel Ramp, and Louise Boronyak. "'Pest' and Resource: A Legal History of Australia's Kangaroos." *Animal Studies Journal* 1, no. 1 (2012): 17–40. https://ro.uow.edu.au/asj/vol1/iss1/3.

Bradley, Daniel G., and David A. Magee. "Genetics and Origins of Domestic Cattle." In *Documenting Domestication: New Genetic and Archaeological Paradigms*, edited by Melinda A. Zeder, Daniel G. Bradley, Eve Emshwiller, and Bruce D. Smith, 317–28. Berkeley: University of California Press, 2006.

Breckwoldt, Roland. *A Very Elegant Animal: The Dingo*. North Ryde: Angus and Robertson, 1988.

Cairns, Kylie M., and Alan N. Wilton. "New Insights on the History of Canids in Oceania Based on Mitochondrial and Nuclear Data." *Genetica* 144, no. 5 (2016): 553–65. https://doi:10.1007/s10709-016-9924-z.

Calaby, J. H., and W. E. Poole. "Keeping Kangaroos in Captivity." *International Zoo Yearbook* 11, no. 1 (1971): 5–12.

Cawte, J. E., Nari Djagamara, and M. G. Barrett. "The Meaning of Subincision of the Urethra to Aboriginal Australians." *British Journal of*

Medical Psychology 39, no. 3 (1966): 245–53.

Chaix, Louis. "A Preboreal Dog from the Northern Alps (Savoie, France)." In *Dogs Through Time: An Archaeological Perspective*, edited by Susan Janet Crockford, 49–59. Oxford: Archaeopress, 2000.

Chapman, Margaret. "Kangaroos and Feral Goats as Economic Resources for Graziers: Some Views from South-West Queensland." *Rangeland Journal* 25, no. 1 (2003): 20–36. https://doi.org/10.1071/RJ03003.

Clutton-Brock, Juliet, ed. *The Walking Larder: Patterns of Domestication, Pastoralism, and Predation*. London: Unwin Hyman, 1989.

Coppinger, Raymond, and Lorna Coppinger. *What Is a Dog?* Chicago: University of Chicago Press, 2016.

Corbett, Laurie. *The Dingo in Australia*. Sydney: University of New South Wales Press, 1995.

Crane, Eva. *The World History of Beekeeping and Honey Hunting*. New York: Routledge, 1999.

Crittenden, Alyssa N. "The Importance of Honey Consumption in Human Evolution." *Food and Foodways* 19, no. 4 (2011): 257–73. https://doi.org/10.1080/07409710.2011.630618.

Croft, David B. "Kangaroos Maligned—Sixteen Million Years of Evolution and Two Centuries of Persecution." In *Kangaroos: Myths and Realities*, 3rd ed., edited by Maryland Wilson and David B. Croft, 17–31. Melbourne: Australian Wildlife Protection Council, 2005.

Davies, S. J. J. F. "The Food of Emus." *Australian Journal of Ecology* 3, no. 4 (1978): 411–22. https://doi.org/10.1111/j.1442-9993.1978.tb01189.x.

———. "The Natural History of the Emu in Comparison with That of Other Ratites." *Proceedings of the International Ornithological Congress* 16 (1976): 109–20.

Davies, S. J. J. F., M. W. R. Beck, and J. P. Kruiskamp. "Results of Banding 154 Emus in Western Australia." *CSIRO Wildlife Research* 16, no. 1 (1971): 77–79.

Delaney, D. A., M. D. Meixner, N. M. Schiff, and W. S. Sheppard. "Genetic Characterization of Commercial Honey Bee (*Hymenoptera: Apidae*) Populations in the United States by Using Mitochondrial and Microsatellite Markers." *Annals of the Entomological Society of America* 102, no. 4 (2009): 666–73.

Diamond, Jared. "The Worst Mistake in the History of the Human Race." *Discover*, May 1987. http://discovermagazine.com/1987/may/02-the-worst-mistake-in-the-history-of-the-human-race.

Dollin, Anne, and Les Dollin. "Tracing Aboriginal Apiculture of Australian Native Bees in the Far North-West." *Australas Beekeeper* 88, no. 6 (1986): 118–22.

Dzoma, B. M., S. Sejoe, and B. V. E. Segwagwe. "Commercial Crocodile Farming in Botswana." *Tropical Animal Health and Production* 40, no. 5 (2008): 377–81. https://doi.org/10.1007/s11250-007-9103-4.

Estrada, Alejandro, Paul A. Garber, Anthony B. Rylands, Christian Roos, Eduardo Fernandez-Duque, Anthony Di Fiore, K. Anne-Isola Nekaris, et al. "Impending Extinction Crisis of the World's Primates: Why Primates Matter." *Science Advances* 3, no. 1 (2017): e1600946. https://doi.org/10.1126/sciadv.1600946.

Fijn, Natasha. "Dog Ears and Tails: Different Relational Ways of Being with Canines in Aboriginal Australia and Mongolia." In *Domestication Gone Wild: Politics and Practices of*

Multispecies Relations, edited by Heather Anne Swanson, Marianne Elisabeth Lien, and Gro B. Ween, 72–93. Durham: Duke University Press, 2018.

———. "Living with Crocodiles: Engagement with a Powerful Reptilian Being." *Animal Studies Journal* 2, no. 2 (2013): 1–27.

———. *Living with Herds: Human-Animal Coexistence in Mongolia*. Cambridge: Cambridge University Press, 2011.

———. "Sugarbag Dreaming: The Significance of Bees to Yolngu in Arnhem Land, Australia." *Humanimalia* 6 (2014): 1–21.

Fijn, Natasha, and Marcus Baynes-Rock. "A Social Ecology of Stingless Bees." *Human Ecology* 46, no. 2 (2018): 207–16.

Fillios, Melanie A., and Paul S. C. Taçon. "Who Let the Dogs In? A Review of the Recent Genetic Evidence for the Introduction of the Dingo to Australia and Implications for the Movement of People." *Journal of Archaeological Science: Reports* 7 (2016): 782–92.

Flanagan, Richard. "Everyone Suffers in the Politics of Hate." *Age*, December 3, 1997.

Francis, Richard C. *Domesticated: Evolution in a Man-Made World*. New York: W. W. Norton, 2015.

Franklin, Adrian. *Animal Nation: The True Story of Animals and Australia*. Sydney: University of New South Wales Press, 2006.

Fuentes, Agustín. *The Creative Spark: How Imagination Made Humans Exceptional*. New York: Penguin Books, 2017.

———. "Monkey and Human Interconnections: The Wild, the Captive, and the In-Between." In *Where the Wild Things Are Now: Domestication Reconsidered*, edited by Rebecca Cassidy and Molly H. Mullin, 123–45. Oxford: Berg, 2007.

Fuentes, Agustín, and Eduardo Kohn. "Two Proposals." *Cambridge Journal of Anthropology* 30, no. 2 (2012): 136–46. https://doi.org/10.3167/ca.2012.300209.

Fuller, Robert S., Michael G. Anderson, Ray P. Norris, and Michelle Trudgett. "The Emu Sky Knowledge of the Kamilaroi and Euahlayi Peoples." *Journal of Astronomical History and Heritage* 17, no. 2 (2014): 1–13.

Galvani, Alison P., and Montgomery Slatkin. "Evaluating Plague and Smallpox as Historical Selective Pressures for the CCR5-Δ32 HIV-Resistance Allele." *Proceedings of the National Academy of Sciences* 100, no. 25 (2003): 15276–79. https://doi.org/10.1073/pnas.2435085100.

Gammage, Bill. *The Biggest Estate on Earth: How Aborigines Made Australia*. Crows Nest, NSW: Allen and Unwin, 2011.

Garde, Murray, ed. *Something About Emus: Bininj Stories from Western Arnhem Land*. Canberra: Aboriginal Studies Press, 2017.

Germonpré, Mietje, Mikhail V. Sablin, Rhiannon E. Stevens, Robert E. M. Hedges, Michael Hofreiter, Mathias Stiller, and Viviane R. Després. "Fossil Dogs and Wolves from Palaeolithic Sites in Belgium, the Ukraine, and Russia: Osteometry, Ancient DNA, and Stable Isotopes." *Journal of Archaeological Science* 36, no. 2 (2009): 473–90. https://doi.org/10.1016/j.jas.2008.09.033.

Gordon, Tulo. *Milbi: Tales from Queensland's Endeavour River*. Canberra: Australian National University Press, 1979.

Graham, Alistair. *Eyelids of Morning: The Mingled Destinies of Crocodiles and*

Men; Being a Description of the Origins, History, and Prospects of Lake Rudolf, Its Peoples, Deserts, Rivers, Mountains, and Weather. New York: New York Graphic Society, 1973.

Greco, Mark K., Dorothee Hoffmann, Anne Dollin, Michael Duncan, Robert Spooner-Hart, and Peter Neumann. "The Alternative Pharaoh Approach: Stingless Bees Mummify Beetle Parasites Alive." *Naturwissenschaften* 97, no. 3 (2010): 319–23. https://doi.org/10.1007/s00114-009-0631-9.

Guintard, Claude. "On the Size of the Ure-ox or Aurochs (*Bos primigenius Bojanus*, 1827)." In *Archäologie und Biologie des Auerochsen*, edited by Gerd C. Weniger, 7–21. Mettmann: Wissenschaftliche Schriften des Neanderthal Museums, 1999.

Guy, Robin D. "The Honey Hunters of Southern Africa." *Bee World* 53, no. 4 (1972): 159–66.

Halcroft, Megan, Robert Spooner-Hart, and Lig Anne Dollin. "Australian Stingless Bees." In *Pot-Honey: A Legacy of Stingless Bees*, edited by Patricia Vit, Silvia R. M. Pedro, and David Roubik, 35–72. New York: Springer, 2013.

Halcroft, Megan T., Robert Spooner-Hart, Anthony M. Haigh, Tim A. Heard, and Anne Dollin. "The Australian Stingless Bee Industry: A Follow-Up Survey, One Decade On." *Journal of Apicultural Research* 52, no. 2 (2013): 1–7. https://doi.org/10.3896/IBRA.1.52.2.01.

Hall, Matthew. "Beyond the Human: Extending Ecological Anarchism." *Environmental Politics* 20, no. 3 (2011): 374–90. https://doi.org/10.1080/09644016.2011.573360.

Hamilton, Annette. "Aboriginal Man's Best Friend?" *Australian Journal of Anthropology* 8, no. 4 (1972): 287–95.

Harshman, John, Edward L. Braun, Michael J. Braun, Christopher J. Huddleston, Rauri C. K. Bowie, Jena L. Chojnowski, Shannon J. Hackett, et al. "Phylogenomic Evidence for Multiple Losses of Flight in Ratite Birds." *Proceedings of the National Academy of Sciences* 105, no. 36 (2008): 13462–67. https://doi.org/10.1073/pnas.0803242105.

Harvey, Fiona, Andrew Wasley, Madlen Davies, and David Child. "Rise of Mega Farms: How the US Model of Intensive Farming Is Invading the World." *Guardian*, July 18, 2017. https://www.theguardian.com/environment/2017/jul/18/rise-of-mega-farms-how-the-us-model-of-intensive-farming-is-invading-the-world?CMP=share_btn_tw.

Heard, Tim. *The Australian Native Bee Book: Keeping Stingless Bee Hives for Pets, Pollination, and Delectable Sugarbag Honey.* West End, Brisbane: Sugarbag Bees, 2015.

Heath, Jeffrey. *Nunggubuyu Myths and Ethnographic Texts.* Canberra: Australian Institute of Aboriginal Studies, 1980.

Hornborg, Alf. "Environmentalism, Ethnicity, and Sacred Places: Reflections on Modernity, Discourse, and Power." *Canadian Review of Sociology / Revue Canadienne de Sociologie* 31, no. 3 (1994): 245–67.

Hossain, Sakhawat, Firoj Jaman, Mushtaq Ahmed, Mokhlesur Rahman, and Saidur Rahman. "High Hatching Success of Saltwater Crocodile (*Crocodylus porosus*) in a Commercial Crocodile Farm of Bangladesh." *University Journal of Zoology, Rajshahi University* 31 (2012): 35–38.

Humphrys, Elizabeth, and Damien Cahill. "How Labour Made Neoliberalism." In *How Labour Built Neoliberalism*, edited by Elizabeth

Humphrys, 167–206. Leiden: Brill, 2018.

Hutton, Jon M., and Grahame J. W. Webb. "The Principles of Farming Crocodiles." In *Proceedings of the Second Regional Meeting (Eastern Asia, Oceania, Australasia) of the Crocodile Special Groups of the Species Survival Commission, IUCN, Convened at Darwin, Northern Territory, Australia*, edited by R. A. Luxmoore, 1–38. Gland, Switzerland: IUCN, 1993.

Ignatieva, Maria, and Marcus Hedblom. "An Alternative Urban Green Carpet." *Science* 362 (October 2018): 148–49. https://doi.org/10.1126/science.aau6974.

Ingold, Tim. *The Perception of the Environment: Essays on Livelihood, Dwelling, and Skill.* London: Routledge, 2000.

Jeengar, Manish Kumar, P. Sravan Kumar, Dinesh Thummuri, Shweta Shrivastava, Lalita Guntuku, Ramakrishna Sistla, and V. G. M. Naidu. "Review on Emu Products for Use as Complementary and Alternative Medicine." *Nutrition* 31, no. 1 (2015): 21–27.

Johnson, Murray. "'Feathered Foes': Soldier Settlers and Western Australia's 'Emu War' of 1932." *Journal of Australian Studies* 30, no. 88 (2006): 147–57.

Johnston, S. D., J. Lever, R. McLeod, M. Oishi, E. Qualischefski, C. Omanga, M. Leitner, et al. "Semen Collection and Seminal Characteristics of the Australian Saltwater Crocodile (*Crocodylus porosus*)." *Aquaculture* 422 (2014): 25–35. https://doi.org/10.1016/j.aquaculture.2013.11.002.

Judt, Tony. *Ill Fares the Land: A Treatise on Our Present Discontents.* London: Penguin Books, 2011.

Kortsch, Susanne, Raul Primicerio, Maria Fossheim, Andrey V. Dolgov, and Michaela Aschan. "Climate Change Alters the Structure of Arctic Marine Food Webs Due to Poleward Shifts of Boreal Generalists." *Proceedings of the Royal Society B: Biological Sciences* 282, no. 1814 (2015): 1–9. https://doi.org/10.1098/rspb.2015.1546.

Kraft, Thomas S., and Vivek V. Venkataraman. "Could Plant Extracts Have Enabled Hominins to Acquire Honey Before the Control of Fire?" *Journal of Human Evolution* 85 (2015): 65–74. https://doi.org/10.1016/j.jhevol.2015.05.010.

Kruuk, Hans. *Hunter and Hunted: Relationships Between Carnivores and People.* Cambridge: Cambridge University Press, 2002.

Larsen, Clark Spencer. "The Agricultural Revolution as Environmental Catastrophe: Implications for Health and Lifestyle in the Holocene." *Quaternary International* 150, no. 1 (2006): 12–20. https://doi.org/10.1016/j.quaint.2006.01.004.

Larson, Greger, and Dorian Q. Fuller. "The Evolution of Animal Domestication." *Annual Review of Ecology, Evolution, and Systematics* 45 (2014): 115–36. https://doi.org/10.1146/annurev-ecolsys-110512-135813.

Leach, Helen M. "Human Domestication Reconsidered." *Current Anthropology* 44, no. 3 (2003): 349–68. https://doi.org/10.1086/368119.

———. "Selection and the Unforeseen Consequences of Domestication." In *Where the Wild Things Are Now: Domestication Reconsidered*, edited by Rebecca Cassidy and Molly H. Mullin, 71–99. Oxford: Berg, 2007.

Leeden, Alex C. van der. "Thundering Gecko and Emu: Mythological Structuring of Nunggubuyu Patrimoieties." In *Australian Aboriginal*

Mythology, edited by Les R. Hiatt, 46–101. Canberra: Australian Institute of Aboriginal Studies, 1975.

Lent, Jeremy. *The Patterning Instinct: A Cultural History of Humanity's Search for Meaning*. Amherst: Prometheus Books, 2017.

Leopold, Aldo. *A Sand County Almanac*. 1949. New York: Oxford University Press, 2001.

Letnic, Mike, Melanie Fillios, and Mathew S. Crowther. "Could Direct Killing by Larger Dingoes Have Caused the Extinction of the Thylacine from Mainland Australia?" *PLOS One* 7, no. 5 (2012): 1–5.

Lien, Marianne E. *Becoming Salmon: Aquaculture and the Domestication of a Fish*. Berkeley: University of California Press, 2015.

Lyons, Mitchell B., Charlotte H. Mills, Christopher E. Gordon, and Mike Letnic. "Linking Trophic Cascades to Changes in Desert Dune Geomorphology Using High-Resolution Drone Data." *Journal of the Royal Society Interface* 15, no. 144 (2018): 1–9. https://doi.org/10.1098/rsif.2018.0327.

Maddock, Kenneth. "The Emu Anomaly." In *Australian Aboriginal Mythology*, edited by Les R. Hiatt, 102–22. Canberra: Australian Institute of Aboriginal Studies, 1975.

Main, George. *Gunderbooka: A "Stone Country" Story*. Kingston, ACT: Resource Policy and Management, 2000.

Malecki, Irek, and Graeme Martin. "Emu Farming—Reproductive Technology." Report for the Rural Industries Research and Development Corporation, March 2000. https://doi.org/10.13140/RG.2.1.2143.6245.

Marshall-Pescini, Sarah, Simona Cafazzo, Zsofia Viranyi, and Friederike Range. "Integrating Social Ecology in Explanations of Wolf-Dog Behavioral Differences." *Current Opinion in Behavioral Sciences* 16 (2017): 80–86. https://doi.org/10.1016/j.cobeha.2017.05.002.

Massaro, Flavia C., Mohammed Katouli, Tanja Grkovic, Hoan Vu, Ronald J. Quinn, Tim A. Heard, Chris Carvalho, Merilyn Manley-Harris, Helen M. Wallace, and Peter Brooks. "Anti-Staphylococcal Activity of C-Methyl Flavanones from Propolis of Australian Stingless Bees (*Tetragonula carbonaria*) and Fruit Resins of *Corymbia torelliana* (Myrtaceae)." *Fitoterapia* 95 (June 2014): 247–57. https://doi.org/10.1016/j.fitote.2014.03.024.

Mathews, Freya. "Planet Beehive." *Australian Humanities Review* 50 (May 2011). http://australianhumanitiesreview.org/2011/05/01/planet-beehive.

Mathews, Janet, comp. *The Opal That Turned into Fire: And Other Stories from the Wangkumara*. Edited by Isobel White. Broome: Magabala Books, 1994.

McCullough, Dale R., and Yvette McCullough. *Kangaroos in Outback Australia: Comparative Ecology and Behavior of Three Coexisting Species*. New York: Columbia University Press, 2000.

McHughen, Alan. *Pandora's Picnic Basket: The Potential and Hazards of Genetically Modified Foods*. London: Oxford University Press, 2000.

McKnight, David. "Sexual Symbolism of Food Among the Wik-Mungkan." *Man* 8, no. 2 (1973): 194–209.

Meggitt, Mervyn J. "The Association Between Australian Aborigines and Dingoes." Department of Anthropology, University of Sydney, 1961.

Mitchell, Thomas L. *Journal of an Expedition into the Interior of Tropical Australia, in Search of a Route from Sydney to the Gulf of Carpentaria*.

London: Longman, Brown, Green and Longmans, 1848.

Monbiot, George. *Feral: Searching for Enchantment on the Frontiers of Rewilding*. London: Penguin Books, 2013.

———. *How Did We Get into This Mess? Politics, Equality, Nature*. London: Verso Books, 2016.

Morey, Darcy F. *Dogs: Domestication and the Development of a Social Bond*. Cambridge: Cambridge University Press, 2010.

Morphy, Howard. "Landscape and the Reproduction of the Ancestral Past." In *The Anthropology of Landscape: Perspectives on Place and Space*, edited by Eric Hirsch and Michael O'Hanlon, 184–209. Oxford: Clarendon Press, 1995.

Murphy, Brett P., and David M. J. S. Bowman. "The Interdependence of Fire, Grass, Kangaroos, and Australian Aborigines: A Case Study from Central Arnhem Land, Northern Australia." *Journal of Biogeography* 34, no. 2 (2007): 237–50. https://doi.org/10.1111/j.1365-2699.2006.01591.x.

National Coroners Information System. "Animal-Related Deaths." NCIS Fact Sheet, March 2011. http://www.ncis.org.au/wp-content/uploads/2017/11/Animal-Related-Deaths.pdf.

Neville-Rolfe, C. W. "Some Birds and Beasts." In *Cassell's Picturesque Australasia*, edited by E. E. Morris, 309–20. London: Cassell, 1889. Facsimile ed., London: Child and Henry, 1978.

New South Wales Government Office of Environment and Heritage. "Brewarrina Aboriginal Fish Traps / Baiame's Ngunnhu." February 25, 2014. http://www.environment.nsw.gov.au/heritageapp/ViewHeritageItemDetails.aspx?ID=5051305.

Nicholls, Jason. "Commercial Emu Raising: Using Cool Climate Forage Based Production Systems." Report for the Rural Industries Research and Development Corporation, December 1998. https://www.agrifutures.com.au/wp-content/uploads/publications/98-147.pdf.

Norris, Ray P., and Duane W. Hamacher. "The Astronomy of Aboriginal Australia." *Proceedings of the International Astronomical Union* 5 (2009): 39–47. https://doi.org/10.1017/S1743921311002122.

Norris, Ray P., and Cilla Norris. *Emu Dreaming: An Introduction to Australian Aboriginal Astronomy*. Sydney: Emu Dreaming, 2009.

Olsen, Sandra L. "Early Horse Domestication on the Eurasian Steppe." In *Documenting Domestication: New Genetic and Archaeological Paradigms*, edited by Melinda A. Zeder, Daniel G. Bradley, Eve Emshwiller, and Bruce D. Smith, 245–69. Berkeley: University of California Press, 2006.

O'Malley, P. J., and John M. Snowden. "Emu Products: Increasing Production and Profitability." Report for the Rural Industries Research and Development Corporation, December 1999. https://www.agrifutures.com.au/wp-content/uploads/publications/99-143.pdf.

Orton, David. "Both Subject and Object: Herding, Inalienability, and Sentient Property in Prehistory." *World Archaeology* 42, no. 2 (2010): 188–200.

Osborne, Michael A. "Acclimatizing the World: A History of the Paradigmatic Colonial Science." *Osiris* 15 (2000): 135–51.

Parker, Merryl Ann. "Bringing the Dingo Home: Discursive Representations of the Dingo by Aboriginal, Colonial, and Contemporary

Australians." PhD diss., University of Tasmania, 2006.
Pascoe, Bruce. *Dark Emu, Black Seeds: Agriculture or Accident?* Broome: Magabala Books, 2014.
Pearce-Duvet, Jessica M. C. "The Origin of Human Pathogens: Evaluating the Role of Agriculture and Domestic Animals in the Evolution of Human Disease." *Biological Reviews* 81, no. 3 (2006): 369–82.
Peck, Jamie, and Adam Tickell. "Neoliberalizing Space." *Antipode* 34, no. 3 (2002): 380–404. https://doi.org/10.1111/1467-8330.00247.
Perry, George H., Nathaniel J. Dominy, Katrina G. Claw, Arthur S. Lee, Heike Fiegler, Richard Redon, John Werner, et al. "Diet and the Evolution of Human Amylase Gene Copy Number Variation." *Nature Genetics* 39, no. 10 (2007): 1256–60. https://doi.org/10.1038/ng2123.
Pickering, Michael. "Garawa Methods of Game Hunting, Preparation, and Cooking." *Records of the South Australian Museum* 26, no. 1 (1992): 9–23.
Pierotti, Raymond John, and Brandy R. Fogg. *The First Domestication: How Wolves and Humans Coevolved.* New Haven: Yale University Press, 2017.
Ploeg, Jan van der, Robert R. Araño, and Merlijn van Weerd. "What Local People Think About Crocodiles: Challenging Environmental Policy Narratives in the Philippines." *Journal of Environment and Development* 20, no. 3 (2011): 303–28.
Plumwood, Val. "Animals and Ecology: Towards a Better Integration." Research School of Social Sciences, Australian National University, 2003. https://core.ac.uk/download/pdf/156617082.pdf.
———. "Being Prey." *Terra Nova* 1, no. 3 (1996): 32–44.
———. "Inequality, Ecojustice, and Ecological Rationality." *Social Philosophy Today* 13 (1998): 75–114. https://doi.org/10.5840/socphiltoday 19981315.
Priddel, D., N. Shepherd, and M. Ellis. "Homing by the Red Kangaroo, *Macropus rufus* (*Marsupialia: Macropodidae*)." *Australian Mammalogy* 2 (1988): 171–72.
Probyn-Rapsey, Fiona S. "Dingoes and Dog-Whistling: A Cultural Politics of Race and Species in Australia." *Animal Studies Journal* 4, no. 2 (2015): 55–77.
Quammen, David. *Monster of God: The Man-Eating Predator in the Jungles of History and the Mind.* New York: W. W. Norton, 2004.
Ripple, William J., Christopher Wolf, Thomas M. Newsome, Mauro Galetti, Mohammed Alamgir, Eileen Crist, Mahmoud I. Mahmoud, et al. "World Scientists' Warning to Humanity: A Second Notice." *BioScience* 67, no. 12 (2017): 1026–28. https://doi.org/10.1093/biosci/bix125.
RIRDC (Rural Industries Research and Development Corporation). "Annual Report, 2007–2008," October 1, 2008. https://www.agrifutures.com.au/product/rirdc-annual-report-2007-2008.
Robin, Libby. "Emu: National Symbols and Ecological Limits." In *Boom and Bust: Bird Stories for a Dry Country*, edited by Libby Robin, Robert Heinsohn, and Leo Joseph, 242–65. Collingwood, Melbourne: CSIRO Publishing, 2009.
Rose, Deborah Bird. *Dingo Makes Us Human: Life and Land in an Aboriginal Australian Culture.* Cambridge: Cambridge University Press, 1992.
———. "An Indigenous Philosophical Ecology: Situating the Human."

Australian Journal of Anthropology 16, no. 3 (2005): 294–305. https://doi.org/10.1111/j.1835-9310.2005.tb00312.x.

———. "Slowly: Writing into the Anthropocene." In "Writing Creates Ecology: Ecology Creates Writing," edited by Martin Harrison, Deborah Bird Rose, Lorraine Shannon, and Kim Satchell. Special issue, *Text* 20 (October 2013): 1–14.

Rose, Deborah Bird, Diana James, and Christine Watson. *Indigenous Kinship with the Natural World in New South Wales.* Hurstville: NSW National Parks and Wildlife Service, 2003.

Sablin, Mikhail V., and Gennady A. Khlopachev. "The Earliest Ice Age Dogs: Evidence from Eliseevichi." *Current Anthropology* 43, no. 5 (2002): 795–99.

Sales, James. "The Emu (*Dromaius novaehollandiae*): A Review of Its Biology and Commercial Products." *Avian and Poultry Biology Reviews* 18, no. 1 (2007): 1–20.

Samiei, Kaveh. "Architecture and Urban Ecosystems: From Segregation to Integration." The Nature of Cities, May 26, 2013. https://www.thenatureofcities.com/2013/05/26/architecture-and-urban-ecosystems-from-segregation-to-integration.

Seymour, Roger S., Christina L. Bennett-Stamper, Sonya D. Johnston, David R. Carrier, and Gordon C. Grigg. "Evidence for Endothermic Ancestors of Crocodiles at the Stem of Archosaur Evolution." *Physiological and Biochemical Zoology* 77, no. 6 (2004): 1051–67.

Shepard, Paul. *The Tender Carnivore and the Sacred Game.* Athens: University of Georgia Press, 1973.

Shepherd, N. C. "The Feasibility of Farming Kangaroos." *Rangeland Journal* 5, no. 1 (1983): 35–44.

Shore, Cris. "Beyond the Multiversity: Neoliberalism and the Rise of the Schizophrenic University." *Social Anthropology* 18, no. 1 (2010): 15–29.

Singley, Blake. "'Hardly Anything Fit for Man to Eat': Food and Colonialism in Australia." *History Australia* 9, no. 3 (2012): 27–42. https://doi.org/10.1080/14490854.2012.11668429.

Smith, Bradley, and Peter Savolainen. "The Origin and Ancestry of the Dingo." In *The Dingo Debate: Origins, Behaviour, and Conservation*, edited by Bradley Smith, 55–79. Clayton South, Melbourne: CSIRO Publishing, 2015.

Smith, Bruce D. "A Cultural Niche Construction Theory of Initial Domestication." *Biological Theory* 6, no. 3 (2011): 260–71. https://doi.org/10.1007/s13752-012-0028-4.

Smith, Nicholas. "The Howl and the Pussy: Feral Cats and Wild Dogs in the Australian Imagination." *Australian Journal of Anthropology* 10, no. 3 (1999): 288–305.

Sorensen, Hayley. "Scientists Using Artificial Insemination to Create Super Crocodiles." *NT News*, May 5, 2016. https://www.ntnews.com.au/news/northern-territory/scientists-using-artificial-insemination-to-create-super-crocodiles/news-story/e146f90626981597e0c0765c3b837265.

Steffen, Will, Jacques Grinevald, Paul Crutzen, and John McNeill. "The Anthropocene: Conceptual and Historical Perspectives." *Philosophical Transactions of the Royal Society A* 369 (March 2011): 842–67. https://doi.org/10.1098/rsta.2010.0327.

Sterelny, Kim, and Trevor Watkins. "Neolithization in Southwest Asia in a Context of Niche Construction Theory." *Cambridge Archaeological*

Journal 25, no. 3 (2015): 673–91. https://doi.org/10.1017/S0959774314000675.

Tisdell, Clement A., and Hemanath Swarna Nantha. "Management, Conservation, and Farming of Saltwater Crocodiles: An Australian Case Study of Sustainable Commercial Use." Economics, Ecology, and the Environment Working Paper no. 126, University of Queensland School of Economics, 2005. https://ideas.repec.org/p/ags/uqseee/55068.html.

Titcomb, Margaret, and Mary Kawena Pukui. *Dog and Man in the Ancient Pacific, with Special Attention to Hawaii.* Bernice P. Bishop Museum Special Publication 59. Honolulu: Star-Bulletin, 1969.

Trigger, David S. "Indigeneity, Ferality, and What 'Belongs' in the Australian Bush: Aboriginal Responses to 'Introduced' Animals and Plants in a Settler-Descendant Society." *Journal of the Royal Anthropological Institute* 14, no. 3 (2008): 628–46. https://doi.org/10.1111/j.1467-9655.2008.00521.x.

Trut, Lyudmila, Irina Oskina, and Anastasiya Kharlamova. "Animal Evolution During Domestication: The Domesticated Fox as a Model." *Bioessays* 31, no. 3 (2009): 349–60. https://doi.org/10.1002/bies.200800070.

Tsing, Anna. "The Global Situation." *Cultural Anthropology* 15, no. 3 (2000): 327–60. https://doi.org/10.1525/can.2000.15.3.327.

Tuinen, Marcel van, Charles G. Sibley, and S. Blair Hedges. "Phylogeny and Biogeography of Ratite Birds Inferred from DNA Sequences of the Mitochondrial Ribosomal Genes." *Molecular Biology and Evolution* 15, no. 4 (1998): 370–76. https://doi.org/10.1093/oxfordjournals.molbev.a025933.

Twain, Mark. *Following the Equator: A Journey Around the World.* Hartford, Conn.: American Publishing, 1897.

Viggers, K. L., and J. P. Hearn. "The Kangaroo Conundrum: Home Range Studies and Implications for Land Management." *Journal of Applied Ecology* 42, no. 1 (2005): 99–107.

Vigne, Jean-Denis. "Early Domestication and Farming: What Should We Know or Do for a Better Understanding?" *Anthropozoologica* 50, no. 2 (2015): 123–50. http://dx.doi.org/10.5252/az2015n2a5.

Vigne, Jean-Denis, Isabelle Carrere, François Briois, and Jean Guilaine. "The Early Process of Mammal Domestication in the Near East: New Evidence from the Pre-Neolithic and Pre-Pottery Neolithic in Cyprus." *Current Anthropology* 52, no. 4 (2011): 255–71.

Vincent, Eve, and Timothy Neale. "Unstable Relations: A Critical Appraisal of Indigeneity and Environmentalism in Contemporary Australia." *Australian Journal of Anthropology* 28, no. 3 (2017): 301–23. https://doi.org/10.1111/taja.12186.

Warale, R. H., H. D. Chauhan, Dilip Parmar, R. C. Kulkarni, A. K. Srivastava, R. B. Makwana, M. M. Pawar, and S. R. Bhagwat. "Emu Farming: An Alternative to Indian Poultry." *Trends in Veterinary and Animal Sciences* 1 (2014): 9–14.

Welch, David. "Beeswax Rock Art in the Kimberley, Western Australia." *Rock Art Research: The Journal of the Australian Rock Art Research Association (AURA)* 12, no. 1 (1995): 23–28.

West, Barbara, and Ben-Xiong Zhou. "Did Chickens Go North? New Evidence for Domestication."

Journal of Archaeological Science 15, no. 5 (1988): 515–33.

Wheeler, Gerald Clair. *The Tribe, and Intertribal Relations in Australia*. London: John Murray, 1910.

Wilkins, Adam S., Richard W. Wrangham, and W. Tecumseh Fitch. "The 'Domestication Syndrome' in Mammals: A Unified Explanation Based on Neural Crest Cell Behavior and Genetics." *Genetics* 197, no. 3 (2014): 795–808.

Williams, Joy. "The Inhumanity of the Animal People." *Harper's Magazine*, August 1997, 60–66.

Williams, Richard J., Eric L. Berlow, Jennifer A. Dunne, Albert-László Barabási, and Neo D. Martinez. "Two Degrees of Separation in Complex Food Webs." *Proceedings of the National Academy of Sciences* 99, no. 20 (2002): 12913–16. https://doi.org/10.1073/pnas.192448799.

Worster, Donald. "Transformations of the Earth: Toward an Agroecological Perspective in History." *Journal of American History* 76, no. 4 (1990): 1087–1106.

Wrangham, Richard W. "Honey and Fire in Human Evolution." In *Casting the Net Wide: Papers in Honor of Glynn Isaac and His Approach to Human Origins Research*, edited by Jeanne Sept and David Pilbeam, 149–67. Oxford: Oxbow Books, 2011.

Young, Diana. "Dingo Scalping and the Frontier Economy in the North-West of South Australia." In *Indigenous Participation in Australian Economies: Historical and Anthropological Perspectives*, edited by Ian Keen, 91–108. Canberra: ANU E Press, 2010.

Zeder, Melinda A. "Archaeological Approaches to Documenting Domestication." In *Documenting Domestication: New Genetic and Archaeological Paradigms*, edited by Melinda A. Zeder, Daniel G. Bradley, Eve Emshwiller, and Bruce D. Smith, 171–80. Berkeley: University of California Press, 2006.

———. "The Domestication of Animals." *Journal of Anthropological Research* 68, no. 2 (2012): 161–90. https://doi.org/10.3998/jar.0521004.0068.201.

———. "Pathways to Animal Domestication." In *Biodiversity in Agriculture: Domestication, Evolution, and Sustainability*, edited by Paul Gepts, Thomas R. Famula, Robert L. Bettinger, Stephen B. Brush, Ardeshir B. Damania, Patrick E. McGuire, and Calvin O. Qualset, 227–59. Cambridge: Cambridge University Press, 2012.

———. "A View from the Zagros: New Perspectives on Livestock Domestication in the Fertile Crescent." In *The First Steps of Animal Domestication*, edited by Jean-Denis Vigne and Daniel Helmer, 125–46. Oxford: Oxbow Books, 2005.

INDEX

Aboriginal Australians, 89, 174, 182
- ancestors, 117, 173
- and dingoes, 29, 31, 32, 35, 174
- art, 60, 116
- ceremonies, 28, 59, 62, 118; increase, 117, 174
- clans, 58, 61, 87, 118–19, 139, 18; associations, 139; restrictions, 181
- communities, 177
- connection to land, 89, 91, 178, 182
- cosmologies, 32, 61, 62, 89, 115, 116, 166
- diets, 140–41, 166
- ecologies, 115, 173, 175
- economies, 63, 140
- estates, 178
- girls, 25, 58, 59
- hunting methods, 118, 139, 141
- initiates, 59, 117
- initiation, 117, 142, 178
- institutionalized violence, 25
- killing of 115, 144
- language map, 177–78
- law, 33
- marginalization of, 136, 147
- marriage rules, 93
- moieties, 61
- movements, 140
- ontologies, 113
- paintings, 62
- people, 24, 143
- practices, 24, 156, 182
- ranger programs, 63
- rangers, 61, 62
- relations with animals, 23, 32
- rituals, 58, 62
- sacred objects, 59
- social obligations, 59
- society, 24–25, 140
- song lines, 62
- songs, 142
- stories, 32, 33, 58, 88–89, 90–93, 115, 116, 133
- storytellers, 33, 89, 91
- technology, 59, 142, 166
- traditions, 173–74
- transubstantiation, 61–62
- worldviews, 23, 24, 58, 62, 152

accidents, road, 136
actinomycosis, 138
aggression, 161
Agricultural Board, Australian 121
agriculturalists
- diets of, 10
- life expectancy, 10
- pathologies of, 10
- quality of life, 11
- stature of, 10, 12

agricultural revolution, 11, 14
Akademgorodok, 19
Akerman, Kim, 59
algae, 117
algal blooms, 23
alienation, 181
Alligator River, 33
alpacas, 14, 136
Altman, John, 118, 140–41
American Emu Association, 121
anarchism, ecological, 171–72
ancestry, 4
animal rights
- discourse, 170
- groups, 147

animals
- agency, 170
- choices, 171
- domesticated, 14 15, 162
- farmed, 174
- feral, 143
- in captivity, 85
- intersubjectivity with, 17, 170–71

animals (*continued*)
  livestock, 17, 22, 28, 60, 136, 147, 158, 163, 167, 171
  native, 41, 176, 181
  ownership of, 16
  products from, 15
  sentience, 17, 170
  spirit, 171
  subjectivities of, 171
  wild, 163
Antarctica, 114
Anthropocene, 4
anthropocentrism, 170
anthropology, 23
  practice, 25
anthroposystems, 21
antibiotics, 168
ants, 183
Apis mellifera, 54, 56
  See also bees, honey
apple, emu, 117
Archie, 130
Arnhem Land, 60, 63, 79, 86, 88, 89, 117, 140
  southeastern, 93
  southern, 114
asses, 14, 161
aurochs, 12, 85
Australia
  Central, 28, 142
  climate, 36
  fauna, 21–22, 162
  outback, 36, 133, 143, 144, 145, 147, 166
  unmaking of, 156
Australian Capital Territory, 157
Australian Dingo Sanctuary, 38
Australian National University, 60
Australian Native Bee Research Centre, 64
Australians, colonial, 23, 24, 25, 34, 83, 84, 88, 121, 136, 137, 164, 174, 177
  economies, 119
  imaginations, 36
  land practices, 132, 143, 178, 180
  mindset, 166
  optimism, 143
  perspectives, 63, 143
  policies, 37
  worldview, 178
australopithecines, 54
Australoplebeia australis, 58

Baiami, 143
bamboo, 56
Bangeran billabong, 90
barramundi, 15, 22, 91, 97
bats, 27
battues, 139
Bauhinia, 59
Baynes-Rock, Leni, 94, 98, 183
bears, brown, 55–56
bee-flies. *See* flies, bee
beekeeping, 64, 75, 101
  Aboriginal modes of, 58, 62
  history of, 55
  modern, 174
  rational, 57
bees, honey, 53, 54, 57
  African, 54, 56
  commoditization of, 58
  evolutionary history, 53
  in Australia, 58
  inebriation, 54
  nests, 56
  population decline, 65
  queens, 57, 58
  quietening, 54
  stings, 54
  swarms, 68
  See also stingless bees
Belyaev, Dmitri, 17–19, 162
Berndt, Catherine, 88, 140
Berndt, Ronald, 33, 88, 140
Bindara, 93–110, 165, 166
  establishment of, 99
  hatching sheds, 102
  income, 99
  location of, 94
  main shed, 107
  ponds system, 100–101, 103
  staff, 94, 100, 166
Bininj, 117–19, 152
biodiversity, 147
birds, 114
  megapode, 98
  species being domesticated, 22
black pack, 47
Blondie, 96
bodies, modification of, 89
Boggy Creek, 79, 80, 165, 182
bora, 117
Botswana, 85
boundaries, 85, 178
  colonial, 178
  crossing, 179
  deconstructing, 180
  national, 7
  primacy of, 179

property, 138, 150, 156
species, 18
Bourke, 143, 144, 158
back o', 147
Breckwoldt, Roland, 43
breeding
plant, 18
unseasonal, 18
Brendon, 127, 128, 129
Brewarrina, 143, 144
Brisbane, 66, 69, 71, 104, 176
Buck, 106
buffalo, 140
bug, assassin, 66
Buka, 97
burning, 141, 177
cultural, 147, 182
lack of, 144
mosaic, 139
regimes, 174
times, 175
burra, 142, 143
See also subincision
Bushmen, 85
butcher shops, 158
butterflies, 98
buzzards, black breasted, 118

Cairns, Hugh, 116
Campion, 120
Canberra, 60, 71
Cape York, western, 58
capital, 164, 166
global flow of, 165, 169, 171
imperatives of, 163
capitalism, 3, 63, 133, 180
See also neoliberalism
capitalists, 180
carbon dioxide, 158
carrot, bush, 117
Casanova, 96, 107
cassowaries, 98, 113
catfish, 91
cats, 15, 21, 34, 40, 42
cattle, 12, 84, 130, 181
domestication of, 14
feral, 140
in Australia, 35, 136, 137
Jersey, 85
Mongolian, 21
See also aurochs
causeways, 93
Cawte, John, 142
Centre for Research on Social Inclusion, 1
Chamberlain, Azaria, 36
charities, 8
chickens, 104, 145
breeding, 66
broiler, 17
carcasses, 96, 97, 104, 105–6, 107, 109
commoditization of, 171
domestication of, 14, 167–68, 168
farming, 9, 122, 168
feeding, 168
free-range, 169
individual, 170
meat, 168; free-range, 168
numbers, 168
origins, 167
predators, 168
vaccinations, 9
vaccines, 167, 168
children, 33
chimpanzees, 54
China, 13
cholesterol, 123, 127, 166
Chris, 145, 146
Christie, 145, 146
cicadas, song of, 175
CITES, 84, 85
civilizations
Indigenous, 143
lost, 183
state, 11
climate change, 6
clique, cultural, 176
Clutton-Brock, Juliet, 15–16
Coalsack Nebula, 116, 122
cognition, 53
colonization, 115, 135, 137, 161, 164, 165
consequences of, 25
dogma of, 178
colonizers, 139, 178
colony collapse disorder, 65, 68
communities, ecological, 24, 173
connectedness
ecological, 6, 24, 85, 135, 152, 169, 174, 175
patterns of, 171, 172
realities of, 182
strands of, 180
connections, 2, 6, 21, 33, 133, 174, 178
ecological, 62, 65, 74, 173, 175
generational, 132
global, 8, 181
life-sustaining, 3, 152
conservation, 22, 24, 36, 84, 85, 95, 109, 137
conservationists, 24, 37

consumers, 152, 16, 171, 177, 181
  chicken, 168
  enslaving, 8
consumption, 13, 137, 172, 182
  honey, 53
Corbett, Laurie, 37
corn. *See* maize
corporations, 7
  unfettered, 8
Crane, Eva, 55, 57
Crittenden, Alyssa, 53, 54
crocodiles, freshwater, 99
crocodiles, saltwater
  abstraction of, 85, 88
  agency, 170
  aggressiveness, 86, 101, 102, 103, 105, 106, 107, 108, 110, 162
  ancestral, 86, 88, 89, 90–92, 93, 173
  as husbands, 92–93
  as nurturers, 86, 89
  as predators, 86, 89–92, 93
  as totems, 86, 87
  as vermin, 84
  baby, 92–93, 97, 103, 104
  behaviour, 95, 101, 105, 109
  biological accounts of, 93
  breeding of, 99, 100, 102, 107, 166
  breeding season, 99
  calling to, 104, 105
  commoditization of, 174
  conservation of, 22, 84, 85, 95, 165, 170
  dangerousness, 95
  domestication of, 162, 165–66
  dutiful, 92
  eggs, 86, 92, 93, 99, 101, 166; clutch sizes, 101, 102, 109; collecting, 84, 85, 102, 107, 174, 175; defense of, 107; hatching, 102; incubating, 84, 102, 103, 174, 175; paternity of, 100
  enclosures, 100, 105, 106
  farming, 15, 84, 93–110, 98, 99, 103, 109, 110, 162, 165, 170
  fecundity, 102
  feeding, 96, 103, 104, 105, 106, 166
  fertility, 102, 105
  fighting, 94, 109
  fillets, tempura battered, 98
  funeral of, 87
  grading, 103, 104
  growth rate, 99, 101, 102, 107, 109, 162
  habitat, 85
  hatchlings, 84, 103
  heartrate, 96
  histories, 107, 109
  inseminating, 108–9, 110, 166
  killing, intraspecific, 97, 99, 100, 101, 102, 107, 110
  killing of, 85, 87, 165
  males, stud, 109
  mating, 97, 99, 100, 101, 102, 110
  meat, 95
  metabolisms, 99
  metamorphosis, 89–92
  nasal dish, 100, 105
  nests, 86
  numbers, 83, 84, 85, 95, 99, 170
  objectification of, 85–86, 170–71
  penises, 108
  personhood, 93
  plush, 94
  poaching, 84
  population, 99
  production, 84, 102
  protection of, 85, 99
  ranching, 22, 84
  range, 99
  recalcitrance, 105, 109
  relations, intraspecies, 97, 99, 101, 105, 106, 107, 109
  relations with humans, 93, 103, 104, 105, 108
  relocation of, 87
  respect for, 173
  scales, 94; belly, 102, 166
  scent glands, 100
  sedation of, 108
  semen, 108, 109; collection, 108; contamination of, 108
  sense perception, 104, 106, 107
  separations of, 100, 101, 110
  sex determination, 102, 103
  singularity of, 85, 86, 165
  sizes, 96, 103, 105
  skins, 83, 84, 95, 165; quality, 102, 103, 107, 109; toughness of, 95; trade in, 84; value of, 83, 95, 99
  skulls, 94
  slaughter, 84, 98, 103
  spiritual power, 98
  stories of, 89–93
  stress, 101, 162
  subjectivities, 170–71
  submissiveness, 107
  super, 108
  tameness, 162
  taxidermed, 94
  teeth, 94
  territoriality, 99–100, 103, 107

traits, heritable, 102
unmaking of, 88, 109–10
vengeful, 92, 93
weight, 85, 170
wild, 83, 84
wild capture, 100
wild release, 84
wives of, 92–93
worth, 83

Crownownership, 16, 138, 146, 156
CSIRO, 119, 137–38

Dalabon, 114–15
Daly River, 91
dancing, 86, 90–91
Darling River, 143
Dawa, 44–49
death, 32
deer, rusa, 98
deforestation, 23
democracy, 67
liberal, 172
Depression, Great, 120
deregulation, 7, 8, 9
Descartes, René, 3
deserts
Central, 88, 116, 116
ecological, 21, 168, 169
Western, 33, 88, 141
destocking, 151
destruction, environmental, 11, 23, 24, 135
detour task, 31
dhumarr, 74
Dhuwa moiety, 61, 118
diabetes, 166
Diamond, Jared, 10, 156, 161
didgeridoo. *See* dhumarr
diets, plant based, 169
Dingo
and Moon, 32
Dingo Conservation Solutions, 46
dingo fence
length, 35
dingoes
aggression, 46, 162
aloofness, 39, 43
archaeological record, 28, 164
as escape artists, 39, 47, 162
as food, 27, 28, 164
as hunting companions, 31
as pets, 22, 29, 34, 41–44
as predators, 34–35, 36, 37, 50–51, 118
attacks on humans, 37, 41
attitudes towards, 41–42
bark-coughs, 45
bonding, 40, 44
bounties on, 35
breeding, 48
breeding season, 38
capacity for reason, 42
captive breeding, 39–40, 51
climbing ability, 50
coloration, 41, 47
conceptualizations of, 36, 37
domestication of, 51, 162, 164, 174
dreaming, 33, 143
farming of, 35
flexibility, 45, 50
genetic diversity of, 27
genetic testing, 38, 47
habituation, 47
howling, 47
hybridisation with dogs, 37, 41, 47, 50, 51, 164
iconicity, 164
in captivity, 38, 42, 50, 51
intractability, 43, 162
introduction to Australia, 27, 164
killing of, 29, 34–35, 43, 49, 164
origins, 164
puppies, 28–29, 33, 39, 42, 51, 97, 174; cooking method, 29
prey drive, 42, 50
recalcitrance, 33, 44
representations of, 36
temperaments, 40, 43
territoriality, 50
unmaking of, 50
whelping season, 28
wild release, 40, 50, 174
disconnection,ecological, 4, 179
diseases, infectious, 10, 11, 12
dispersal
inhibited, 31–32, 44
Djandomerr, George, 119
Djandomerr, Jack, 118
Djinan-derara, 91
DNA, 4–5, 47, 164
Doc, 126, 128
Dog, 140
doggers, 35
dogs, 30, 164, 174
aggressiveness, 29
and Aboriginal people, 30
as food, 13
Australian cattle, 34
blind loyalty, 44
breeding, 38

dogs (*continued*)
breeds, 13
domestication of, 13, 31–32
emu finding, 131
evolution of, 13, 30
going bush, 51
hunting, 141
in stories, 91
semi-domesticated, 28
show, 40
submissiveness, 44
wild, 130–31, 143 (*see also* dingoes)
working, 43
domestication
and hunting, 14, 31
and predators, 22
as pet-keeping, 15
as separation, 21, 78, 110, 163, 175
as unmaking, 10, 133
consequences of, 10–12, 14, 163
cutting-edge, 168
definitions of, 15–17, 21, 110, 171
history of, 10, 23
in Australia, 21
new wave of, 25, 163, 168
nonhierarchical, 171
rejection of, 171
reverse, 28
suitability for, 136, 156, 162
traits, 18
Western, 62, 74
See also tameness
domestication superhighway, 14
dreaming
beings, 33, 91, 140, 142
stories, 114, 116, 118, 133, 140, 142
dualism, 3, 6, 25
good/evil, 167, 182
human/nature, 8, 9, 152
mind/body, 182
native/introduced, 167
nature/culture, 63, 167, 182
pure/hybrid, 167
reason/emotion, 167
soul/body, 182
wild/domestic, 43, 167
dualisms, 28, 167, 182
toxic, 182
ducks, 14, 90
Ducos, Pierre, 16
Durack, Michael, 28
dust, interstellar, 116

eagles, wedge-tailed, 50, 131
echidnas, 136
ecoanarchism, 172
ecologies, 74
Australian, 146, 174
breakdown of, 143
complexity of, 6, 8, 21, 62, 110, 169, 173, 182
costs to, 8
diverse, 179
healthy, 24, 74, 156, 178, 179, 182
local, 74, 172, 174, 180
simplified, 180
threatening, 74
traditional, 162
unmaking of, 63, 158
econationalism, 36, 165
See also nativeness
economy
Australian, 35, 63, 132, 136
knowledge, 1
ecotones, 139
education systems, 5
eduction, 73
Egypt, ancient, 55
Ekebergia capensis, 56
electroejaculation, 108
elephant birds, 113
empathy, 64–65
emptiness, psychological, 181
Emu Bliss, 123–29, 130
products, 126
emus
aggression, 127, 129, 133
ancestors of, 113, 114
as ancestors, 118
as persons, 119
behaviour, 121, 127
body fat, 115, 122, 125, 127, 133
breeding, 121, 128, 129, 132, 133
capture, 119, 121
chicks, 121, 124, 131
claws, 122, 129
commoditization of, 121
cooking, 115
curiosity, 118, 125, 129, 130
dark, 116
declawing, 133
diets, 117–18, 121
diseases, 133
domestication of, 132, 166
ecologies, 116, 122, 132
eggs, 113, 115, 130; collecting, 117, 127, 128, 129, 131, 132; incubating, 119, 128, 131, 132; predation on, 130–31;

protection of, 162; selection of, 128; uses, 122, 128; whitened, 129
engraving of, 116
eradication, 120
farmed, 15, 22
farmers, 132, 133, 162
farming, 121–22, 123–33, 156, 180; scalability of, 166
feathers, 122, 131, 133
feeding, 125, 127, 129, 133, 139
flightlessness, 113, 119, 121
food preferences, 117
guts, 127
habits, 117; feeding, 117, 118, 119
hunting of, 118–19
inseminating, 133, 166
in zoos, 125
killing of, 120
male, 124, 132–33
management, 130
mating, 117, 125, 127, 130, 131, 133
meat, 15, 122, 132, 133; distribution of, 118
migratory routes, 119, 166
movements, 117, 119, 126, 127, 128, 130, 132, 180
nesting, 118, 131, 132
numbers, 121, 122
oil, 127, 132, 133; therapeutic properties, 122–23, 124, 129, 166
plush, 123
poo, 118
predators, 118, 125, 130, 131, 132, 163, 166, 171
products, 122
relations with humans, 125, 128, 130
running speed, 126, 129
size, 113
skinning, 115
skins, 122, 133
sky, 116–17, 133
slaughter, 121
squabbling, 127
starving, 123
stone arrangement, 117
stone swallowing, 115
stories, 114–15, 117, 118
swimming ability, 114, 115, 118
tameness, 162
territoriality, 119, 126, 127, 128, 133
toughness, 120, 121
unmaking of, 122, 125, 132, 133, 166
vaccinating, 133
value, 122
wanderlust, 119, 128, 130
watering, 125, 126
weight, 121, 122
wild, 122, 123, 131, 132
worming, 133
wounded, 121
Emu War, 120–21
Endeavour River, 92
England, meadows of, 143
Ethiopia, 15–16, 56, 83
security situation in, 22
ethnography, 115
methods, 25
See also anthropology
Europeans, 12, 36, 37, 54, 136, 139, 143, 181
arrival of, 30
evolution
backwards, 28
human, 30, 53–54, 172
experiments
behavioural, 31
extinctions, 6, 23, 27, 37, 65, 95, 137, 144, 163, 164
extraction, 172

farmers, 18, 135, 147, 148, 149, 150, 156, 177
ancient, 10
colonial, 176, 177
Neolithic, 23
Oromo, 56
Tasmanian, 36
farm-fox experiment, 17–20, 31, 162
farming
Australian concept of, 176
colonial, 167, 176, 177
early, 163
factory, 152, 170
fire-stick, 139, 156, 178, 181
industrial, 3, 152
methods, 9, 174
mixed, 3, 120
monocultures, 3
organic, 182
transition to, 10
fashion, 83, 165
Fat Bill, 96–97
fence posts, 100, 151
fence, rabbit-proof, 35–36, 119
fencing, 98, 99, 100, 104, 105, 110, 125–26, 128, 130, 147
absence of, 16
chain-link, 94, 103, 106
cost of, 138, 146, 150
electric, 47, 132
high, 162, 174

fencing (*continued*)
  inadequacy of, 151
  internal, 110
  mesh, 126, 132
Fertile Crescent, 14
figs, 175
Fijn, Natasha, 60–63, 86
  fieldwork, 60, 86
  on domestication, 17
fire, 139, 144, 158, 159, 179
  control of, 54
  drives, 141
  hiding of, 89
First World War, 120
fish, 91, 114
  catching, 140, 141
  farmed, 15, 20, 21, 22
fishing industry, 98
flavonoids, 65
flea collars, 51, 174
flies
  bee, 177
  March, 175
  phorid, 69, 174
  syrphid, 73, 76, 77, 78, 79, 80, 174; appearance, 73, 76; life cycle, 74, 76; Yolngu conceptions of, 74
flights, orienting, 75
flowers, 65, 69
  cadaghi, 71
  grevillea, 117
  holly-leaved pea, 117
  silky oak, 175
  wattle, 117
foods, bush, 141, 177
food webs, 6, 171, 172, 174, 175, 180, 181
Fords Bridge, 147, 179
Forestry Commission, 182
forests, 44–45, 49, 61, 69
  fruits of, 63
  homogeneity of, 70
  rain, 71
  scrub, 60, 143, 148
foxes, 124, 131
  aggressiveness, 18
  as pets, 19–20
  ecologies, 20
  fearfulness, 18
  habitats, 20
  nurture of, 19
  selective breeding of, 18
  tail-wagging, 19
Francis, Richard, 28
frogs, 164
fruit
  blue tongue, 117
  red leea, 117
Fuentes, Agustín, 21, 22
fungal infections, 73, 103
fungi, 73, 74, 110
fungicides, 102

Gage, Clinton, 37
Galton, Francis, 31
Gammage, Bill, 139
Ganhaarr, 92
Garawa, western, 141
gardens, community, 182
Garland, Fiona, 146
gastroliths, 115
gazelles, 161
genetic engineering, 169
genitalia, nonspecific, 58
Gladstone, 126
globalization, 8–9, 165, 174
goannas, 89, 130–31, 132
  as food, 33
  as prey, 49, 51
  stories of, 90
goats, 45
  abundance of, 149
  ancestors of, 161
  as pests, 149
  domestication of, 14
  farming, 150, 156
  feral, 149
  harvesting, 146
  kids, 163
  meat, 158
  movements, 150, 156
  mustering, 149
  ownership of, 156
  profitability of, 146, 149, 150
  wild, 163
Gondwanaland, 113–14
Gordon, Tulo, 92
Goulburn Islands, 89, 90
government, 177
  Australian, 9, 71, 120, 137
  failures, 8
  regulations, 168
  responsibilities, 8
grass, 128, 136, 139, 143, 159
  eaters, 34

unmaking of, 154
See also pasture
Greece
ancient, 3, 6
prehistoric, 10
Gucci, 101
Gulf of Carpentaria, 139
Gumatj
elders, 87
land, 87
Gumbaynggirr language, 45
Gundabooka National Park, 145
guns, 35, 37, 83, 87–88, 120, 140, 141, 151–52, 153
machine, 120, 121
Gunwinggu, 140–41
Guringai, 115–16
legacy, 116
Gwunbiuribiri, 90

Hall, Matthew, 171
Hamilton, Annette, 29–30
handbags, 83, 94, 95, 102, 165
hare-wallaby, eastern, 137
harms, ecological, 172, 179
Hayek, Friedrich, 7
Heard, Tim, 64–71, 72, 75, 76, 77, 80
Heath, Jeffrey, 93
heather, coastal, 154, 155
Helen, 148–50
herbicides, 72, 169
hierarchies
deconstruction of, 175
economic, 172
social, 171, 172
highways, 123
hive beetles, 73, 74, 174
larvae, 73, 80, 182
hive boxes, 79
hives
ceramic, 55
construction of, 56, 71
death of, 72, 183
designs, 57
differentiating, 69
fumigation of, 56
ghost, 183
isolation of, 74
Langstroth, 57–58
ownership of, 55
production, 66
protection of, 56, 64, 75, 76, 78
salvaging, 79–80, 183
splitting, 66–67, 68, 72, 175
swapping, 70–71
transportation of, 75, 78
hive tools, 66, 79
Homo erectus, 54
Homo habilis, 53
honey
consumption of, 53, 54, 62; prohibitions on, 58, 59
containers, 55, 59, 61, 62
giving of, 59
hunters and hunting, 54, 55, 62, 63
importance of: in evolution, 53–54; in social relations, 58–59
production, 68
tools for extraction, 59, 61 (*see also* hive tools)
honey badgers. *See* ratels
hormones, 19
Hornborg, Alf, 109
horses
domestication of, 14
in Ethiopia, 15–16, 21
size variation among, 12
humanism, 8, 162
humans
ancestral, 5, 10, 53, 54, 92, 152
autonomy, 24
communities, 162
cooperating, 5
dominance over animals, 170, 171
niche, 25
numbers, 169
territoriality, 55
hunter-gatherers, 171
ancient, 10 (*see also* humans: ancestral)
and dogs, 13
average height of, 10
diets of, 10
mobility, 11
hunting
and domestication, 31
and human evolution, 30–31
strategies, 14
hybridity, 37–38
hyenas, 1, 22, 106

imagination, colonial, 36
imperialism, 135
imprinting, 125
inbreeding, 31

incest, 60
India, 133, 165, 169
Indigenous Philosophical Ecology, 24, 86
  as alternative to unmaking, 169–75
  prescriptions of, 175
  See also reconnection
individualism, 3, 7–8, 169
Indo-Europeans, proto, 3
industrialization, 8, 23
inequality, 7, 172
  economic, 180
  global, 11
  social, 11, 179, 180
insemination, artificial, 40, 108, 133
insurance companies, 144–45, 154
integrity, ecological, 87
interdependencies, 74, 139, 144, 169, 175
intervention, human, 69
introduced species, 24, 71, 135, 144, 156, 165

Jankuntjara, 29–30

Kalenjin, 54
Kamilaroi, 116, 117
Kangaroo, 140, 142
kangaroo industry, 136, 137, 147, 153
kangaroos, 114
  abundance, 136, 153, 158, 181
  aggressiveness, 163
  ancestral, 142, 143
  as food, 136, 137
  as functionaries, 140
  as pets, 22, 145, 148
  as protagonists, 140
  as vermin, 136–37, 143, 144, 146
  attacks on humans, 154
  bounties on, 35, 137
  carcasses, 137
  diseases, 138 (*see also* actinomycosis)
  domestic, 146
  dreaming, 140
  environmental credentials, 147
  farming, 137, 139, 146, 147, 167; context of, 163, 166; viability of, 138, 156, 158, 159
  feeding, 145
  females, 137, 138, 145
  fighting, 138
  flight distance, 151
  fostering, 138
  grey, 157
  growth rate, 138
  habits, 136
  harvesting: commercially, 148, 150; male only, 138
  haunch of, 136
  hill, 141
  home range sizes, 157
  hunting of, 136, 139, 141, 151, 156
  joeys: dying, 145; orphaned, 145, 152, 153
  killing, 137, 144, 152, 153, 158; permits for, 146–47, 148
  kinship with, 142–43
  males, 137, 138, 151
  management, 146
  meat, 136, 138, 140, 151, 166, 167; availability of, 158; price of, 146, 149, 150, 156; processing, 148, 158, 180–81; processors, 137; trade in, 137, 138; wasting, 148
  movements, 138, 139, 146, 147, 156–58
  naming, 145–46
  numbers, 151
  perceptions of, 147
  photos of, *134*, 153, 155, 181
  physiology, 142
  populations, 137
  pouch bacteria, 137, 138, 149
  pouch young, 137, 144
  red, 157
  reproduction, 138
  resilience, 137
  roadkill, 144, 158
  site fidelity, 146, 157–58
  skinning, 137
  skins, 136, 137
  spearing, 139, 141
  steaks, 158
  stories of, 140
  suburban, 154, 155
  tail, 114
  tameness, 146, 155, 156, 163
  veterinary treatment, 167
  welfare, 150, 151
Karen (tour guide), 94–98, 104
Karlaya, 178
Kemp, Ross, 150–52
Kennedy, Tye, 176
K'Gari (Fraser Island), 36–37
kidnapping, 92–93
Kikuyu, 54
Kimberley, 28, 59, 60, 88
Kinchega National Park, 157
King River, 90

King Wally, 100
kinship, 159
    interspecies, 24, 142–43, 173–74
kiwis, 113
Kluskap's Mountain, 2–3
kookaburras, 45, 183
Kubarkku, Mick, 118
Ku-ring-gai Chase, 115, 116

labor
    divisions of, 11
    market, 7
lactose tolerance, 12
land, 145
    accumulation of, 179–80
    commoditization of, 179
    control of, 179–80
    dependence on, 181
    depersonalization of, 179
    fertile, 143
    holistic approach to, 179
    integrity of, 156
    management practices, 179
    ownership of, 159, 177, 178, 179; restricting, 180
    private, 144
    responsibilities toward, 179
    trust in, 181
Landcare, 147, 149, 179
landholders. *See* farmers
landscapes
    and stories, 33, 140
    complex, 181
    domesticated, 180
    dreaming, 173
    healthy, 147, 181
    herding, 171
    modern, 173
    relation to living, 33
    unfamiliar, 25
Langstroth, Reverend L. L., 57
late modernity, 2, 3, 58, 85, 109, 133, 158, 170, 175, 181
    and reconnection, 175
    structures of, 171
lawns, 154
Layne, 103–4
Leach, Helen, 163
Le Mont Saint-Michel, 93
Lent, Jeremy, 5–6
Leopold, Aldo, 3
level playing field, 4
Lien, Marianne, 104
lights, solar, 132
Little Black Bee, 176
Liverpool River, 140
lizards, 114, 176
    monitor, 139
llamas, 14, 136
Local Land Services, 146
Look at Me Now Headland, 154–55
Lorenz, Konrad, 43–44
Luke, 104–7
lumpy jaw. *See* actinomycosis
Lysenko, Trofim, 18

Maburrinj Estate, 117
macropods, 106
Maddock, Kenneth, 114–15
maggots, 80, 86
maize, 12, 69
Maluwau'wau, Mathew, 90
mammals, 22, 93, 106, 140–41, 157
    placental, 27
    small, 59
Maragar, Nipper, 33
Margi, 44–51
Maringa Bambu, 116
marriage
    interspecies, 92–93
    rules of, 93
Marshall Thomas, Elizabeth, 42
Marsupial Destruction Act, 35
Marsupial Mouse, 142
marsupials, 114, 138, 139, 142
masseuse, digital, 108
Matthew, 105
measles, 10
media
    social, 41–42
    television, 41–42
Melbourne, 167
Meredith, Maj. G. P. W. 120
militarism, 178
milk, consumption of, 12
Milky Way, 116
Mike (Bindara's owner), 98–102, 107
    education, 98
    employment, 98
    love of crocodiles, 98
    teaching background, 98, 99
Mi'kmaq, 2–3
millet, native, 143
mining, 2–3, 87, 94

Mitchell, Maj. Thomas Livingstone, 139, 143, 144
mites, varroa, 65
moas, 113
Mojo, 44, 45, 47, 48, 49
Monbiot, George, 7, 180
Mongolia, 17, 60
monocultures, 3, 169, 180
Mont Pelerin Society, 7
Moon, Dingo and, 32
morality, 24, 173
Moree, 158
Morgan, Nyerri, 178
Morphy, Howard, 87
moths, 58
  Quandong, 176
  wax, 56
Murchison Ranges, 116
mussels, 92

Nabarlambarl, Peter, 118
Namiyadjad, Peter, 90
Nasanov pheromone, 57
Nathan, 97
national parks, 85
National Parks and Wildlife Service, 36–37, 116, 131, 138, 155, 157
nativeness, 9–10, 64, 164
  See also econationalism
nature, 36
Nawilil, Jack, 118
neoliberalism, 7–9, 180, 181
Neolithic, 12, 17, 163
  revolution, 23
neoteny, 44
Ne-user-re, 55
New England Tablelands, 44
New South Wales
  government, 148
  Governor of, 34
  mid north coast, 153–54
  northern, 25, 71, 131, 143
  northwest, 116
  outback, 144
  rangelands, 147, 153, 154, 156, 158
  suburban, 75
New York Stock Exchange, 8
Ngarinyin, 59
Ngindyal, 116
Ngyamba, 143, 173
Nhulunbuy, 87
niches
  as hard taskmasters, 20
  domestic, 21
  ecological, 167
niche construction, 2, 21
nodes, ecological, 178
Northern Territory, 29, 116, 140
  government of, 84
  See also Arnhem Land
nurture, 19, 30, 31, 164, 174

Oakman, Barry, 42
ochre, 29, 59, 91, 118, 141
Olea Africana, 56
onagers, 161–62
ontologies
  Aboriginal, 23
  Western, 23, 181
Oromo
  agro-pastoralists, 22
  horses, 15–16
  people, 15–16
Orton, David, 17
Osborne, Michael, 135
Oscar, 106
ostriches, 113, 114

panic, metaphysical, 37
Papua New Guinea, 94, 98, 99
Parker, Merryl, 33, 36, 37
pastoralists, 34, 43
  Mongolian, 17, 171, 172
pasture, 44, 126, 130, 151, 154, 158
  competition for, 148, 149
  lush, 139, 143
Pastures and Stock Protection Act, 144
pathogens, 65, 168
pathways, ancestral, 33
patterning instinct, 5
patterns, ecological, 175
people, bird, 90
permaculture, 182
Perth, 176
pesticides, 168, 169
pets, 163, 174
  trade in, 15, 38
pheromones, 100, 176
  See also Nasanov pheromone
Phillip, Arthur, 34
philosophy
  Enlightenment, 6, 135
  Greek, 3, 6
phorid fly, 69, 174
Pickering, Michael, 141
pigs, 14, 140
Pippi Longstocking, 46–51
  brother of, 48
  condition at capture, 46

genetic test, 50
escape, 50
mother of, 48
visit to the vet, 50
Pitjantjatjara, 28
plant extracts, 55
plants
fruiting, 139
native, 176
Plato, 6
plums
black, 117, 118
bush, 158
green, 117
Plumwood, Val, 9, 86, 172, 179
pollution, industrial, 23
polyethism, age, 64
population
densities, 10
growth, 163
populism, 8
Port of Bourke Hotel, 146
possums, 59, 114, 136
Powell, Nick, 77
prefrontal cortex, 5
pregnancy, 59
privatization, 7
Probyn-Rapsey, Fiona, 37
propolis, 64, 79, 80, 174
in art, 60
labyrinth of, 183
production, 65, 68
significance of, 60
toothpaste, 65
purity, 37–38, 164
genetic, 49, 51
See also dingoes: hybridization with dogs

Quammen, David, 84
Queensland, 36, 99
government of, 35, 37, 84
North, 71
rural, 95
South East, 153
Queenslanders, 84

rabbits, 14
control measures, 35
racism, 37
rainbow serpents, 34
ranching, fire-stick. *See* farming: fire-stick
Randell, Capt. William, 143
ratels, 55, 57
rationality, ecological, 172
ratites, 113–14
Reagan, Ronald, 7
real estate
agents, 153
listings, 153
See also land: commoditization of
rebirth, 32
reconnection, 170–76, 179
remoteness
ecological, 9, 172
hierarchies of, 179
reptiles, species being domesticated, 22
resins, 75, 165
cadaghi, 71, 72
tree, 60, 63, 65, 69, 73
resource extraction, 8
restaurants, 95, 98, 158
rewilding, 180
rheas, 113, 114
rifles. *See* guns
RIRDC, 9, 121
Robbie, 105, 107–8
Robin, Libby, 119, 132
Rocksteady, 106
rodents, 27
roo bars, 144
Rose, Deborah Bird, 1–2, 4, 178
fieldwork, 24, 32
metaphysics, 24, 86, 169, 175
Ross, 130
Roundup, 72
See also herbicides
Royal Australian Artillery, 120, 127
royalty, English, 136

salinity, 23
salmon, 104
sauce, bush-plum, 158
scarification, 94
Science, 6, 135
and religion, 6
methods, 31
Western, 113
Second World War, 7, 36, 83
sedentism, 163
selection
destabilizing, 20
differential, 67
for tameness, 19
operational, 20
separation, 100, 101, 169
degrees of, 168

separation (*continued*)
ecological, 20, 28, 51, 55, 72, 78, 132, 158, 162, 166, 181
Seventh Day Adventism, 36
Shane, 124–29, 166
appearance, 124
research work, 124, 127
sheep, 12–13, 148
addiction to, 144
ancestors of, 161
as prey, 34, 130
comparisons with kangaroos, 138
domestication of, 14
farming, 35, 36, 37, 136, 137, 150, 156, 158
Shepard, Paul, 20, 152, 169, 173
shoes, 102
shooters
kangaroo, 137, 146, 148, 150, 152, 153, 156
sporting, 152
sickness, motion, 46
slavery, 11, 180
smallpox, 10, 12, 115
snakes, 45, 51, 89, 104, 118, 130–31, 132
Snappy Tom, 105
socialization, 20, 31, 51, 162
societies
Aboriginal, 24–25
acclimatization, 135–36
agrarian, 11
colonial Australian, 37
ecologically healthy, 173
egalitarian, 171, 172
human, 17, 23, 24–25, 28, 172
illuminated, 133
modern, 11
no such thing as, 3
organization of, 11, 12, 169
settler, 115
state-based, 172
tribal, 171, 172
wider Australian, 63
soil, 181
acidic, 36
degradation, 169
erosion, 23
fertility, 139
healthy, 182
pollution of, 169
South Australia, 35
Southeast Asia, 28
Southern Cross, 116
souvenirs, 94
spears and spearing, 29, 92, 93, 118, 139, 141
species
conservation of, 85
invasive, 72, 148
spiders, golden orb, 176
starch hydrolysis, 12
Steve, 148, 149
stingless bees, 22, 54, 58, 60, 114, 176
and Aboriginal Australians, 58
aroma from, 61, 76–77
as pets, 165
bites from, 58, 67, 71
breeding, 65
brood structures, 67, 73, 79
cognition, 69, 70
commoditization of, 64
defensive reactions, 67, 70
domestication of, 65, 70, 78, 79, 80, 164, 165
drifting, 70
enemies, 66, 69, 76, 162, 174
evolution, 67
feeder designs, 77
feeding, 77–78
fighting, 69, 70–71
flight paths, 177
flight ranges, 68
foraging, 64, 69, 70, 75, 80
hives, 68, 76, 165
hive volumes, 64
homing success, 70
honey from, 62, 63, 65, 67, 74, 165, 174; antimicrobial properties, 65; harvesting, 75, 79
in backyards, 63, 66, 67, 69, 75–80, 164, 174
independence, 68
in Ethiopia, 56
keeping of, 64, 165
nests, 79
nest structures, 58–59, 61, 62, 67
ownership of, 165
pollination, 63–64, 68, 70, 165
protection of, 78, 80, 162
queen cells, 67
queens, 67, 73, 79; production of, 67, 73, 79; replacement of, 101; selection of, 67
reproduction, 64, 67, 68
rescuing, 79, 164
resilience, 72
resistance to pathogens, 65

selection, 65–66, 67
shape recognition, 70
social mobility, 67
sound of, 61
suburban, 64, 66, 67, 68, 70, 162, 165
swarms, 69, 70
tameness, 162
unmaking of, 63, 64, 66, 74, 80
wild, 68, 69, 70
stolen generation, 37
structuralism, 115
Sturt, Charles, 143
subincision, 141–43
sugarbag, 62, 140, 141, 169
ceremony, 74
complex, 74
Sugarbag Dreaming, 60–62
Sulawesi, 27
sun bears, Malayan, 55
superphosphates, 138
supplements, vitamin, 105
supply chains, 152
sustainability, 179
sustainable use, 98, 99
principle of, 84
shift to, 85
Sydney Dingo Rescue, 41
systems, ecological, 85, 109, 136, 139

Talon, 49–50
tameness, 18, 161, 162
and domestication, 162
testing of, 18
tape, electrified, 45, 124
See also fencing: electric
Tara, 71–74, 77
Tasmania, 36
Tasmanian Devil, 27
Tatars, Cheremis, 56
Tetragonula carbonaria, 67, 71
Tetragonula hockingsi, 66
ticks, 46, 50
tests, genetic, 4, 38
Thatcher, Margaret, 3, 7
thiabendazole, 102
Thornton, Gwen, 38–40
thylacines, 27, 36, 164
time
linear, 175
radial, 175
tinamous, 114
Tindale, Norman, 28, 177–78
tools
categories of, 59
hive, 54, 66
making, 54
modifying, 54
stone, 53, 60
totemism, 143, 173
tourism, 36–37, 84, 109
tourists, 36–37, 84, 85, 94, 97, 99, 124, 130, 145, 154
tractors, 127, 129
transnationalism, 7, 8
transubstantiation, 61–62, 173
traps, fish, 91, 143
trees, 44, 45
avocado, 130
banksia, 115, 155
bauhinia, 59
black plum, 118
bloodwood, 91
cadaghi, 71–72
eucalyptus, 126
ironwood, 89, 90
jacaranda, 130
modifying, 56
pecan, 130
Poinciana, 130
Quandong, 176
quinine, 117
sapote, 183
scribbly gum, 115
shade, 70
silky oak
stringybark, 74
stunted, 60
truck drivers, 145
Trut, Lyudmila, 19
turtles, 175

Underhill, Richard, 56–57
universities, 1, 60, 86, 153
university zombies, 2
unmaking, 3–6, 9, 66, 88, 109, 135, 154, 167
apotheosis of, 168
beyond, 182
consequences of, 169
definition of, 4
dominant system of, 182
ingrained, 171
institutions of, 175
legacy of, 169
metaphors of, 182

unmaking (*continued*)
toward, 80
way of, 178
writ large, 133
urbanites, 37

van Halen, Eddie, 124
Vedda, 54
veganism, 152
Vegemite, 64
Victoria
government of, 38, 136
west-central, 116
Victoria River Region, 24
Vinča culture, 17
vinegar, apple cider, 73
viruses, 4
globe-trotting, 133
See also smallpox
vulva envy, 142

Wadedi, Fred, 89
Walbiri, 142
Walgoolan, 120
Walker, Terry, 131–32, 166
wallabies, 48, 49, 50, 91, 106, 136, 137
See also macropods
Wandjili, 90
Wangkumara, 116, 143
warfare, 11
water weed, 117
Watson, Barry, 130
Wayiliwan, 143
welfare, animal, 137, 170
Western Australia, 119, 121
government of, 84, 120
Western Desert, 33
West Shewa, 56
whack-a-mole, anarchist, 172
wheat, 12
farmers, 120
production, 120
Wik-Mungkan, 58
wildness, 48–49
Wilson, Edward, 136
wolves, 13, 20
and humans, 14
in Mongolia, 17, 171
loyalty, 43
pups, 44
women, egg-laying, 92–93
woody weed, 144, 148
wool, 143, 144
Woolgoolga, 153, 158
golf club, 154
soccer pitch, 154
World Health Organization, 168
worms and worming, 46, 51, 98
Worora, 59
Wrangham, Richard, 54
Wunambal, 59
Wunungmirra, Yumitjin (Jimmy), 61–62

Yalangbara, 87
Yarralin people, 24
Yathong Nature Reserve, 157
Yibarbuk, Dean, 117
Yindi, 44–50
Yingi, 90–91
Yirritja moiety, 87, 88, 118
Yolngu, 78, 79
connection to land, 87, 176
ecologies, 86, 88
finding honey, 60, 63
names for sugarbag complex, 61
ontologies, 63, 65, 86, 88
philosophy, 63, 74, 165
relations with bees, 60
relations with crocodiles, 86, 88
worldviews, 23, 62, 63, 173
Yuin, 173

zebras, 161
Zeder, Melinda, 14, 16, 156
Ziggy, 105–6
zombies, 129
university, 2
zoological gardens, 135
See also zoos
zoonoses, 21
zoos, 38, 39, 51, 85, 107, 125